# LACUNARY POLYNOMIALS OVER FINITE FIELDS

# Lacunary Polynomials over Finite Fields

by

L. RÉDEI
*Member of the Hungarian Academy of Sciences*



N·H
P C

1973



NORTH-HOLLAND PUBLISHING COMPANY — AMSTERDAM · LONDON
AMERICAN ELSEVIER PUBLISHING COMPANY, INC. — NEW YORK

Library of Congress Catalog Card Number: 72-88575
North-Holland Publishing Company ISBN: 0 7204 2050 4
American Elsevier Publishing Company ISBN: 0 444 10400 3

Translation of

LÜCKENHAFTE POLYNOME ÜBER ENDLICHEN KÖRPERN

VEB Deutscher Verlag der Wissenschaften, Berlin — Birkhäuser Verlag, Basel
Akadémiai Kiadó, Budapest

Publishers:

NORTH-HOLLAND PUBLISHING COMPANY, AMSTERDAM
NORTH-HOLLAND PUBLISHING COMPANY, LTD., LONDON
AKADÉMIAI KIADÓ, BUDAPEST

Sole distributors for the U.S.A. and Canada:

AMERICAN ELSEVIER PUBLISHING COMPANY, INC.
52 Vanderbilt Avenue
New York, N.Y. 10017

English translation by

I. FÖLDES

English translation edited by
Minerva Translations Ltd., London

Printed in Hungary

*To my Friends, György Hajós, László Kalmár
and Pál Turán*

# CONTENTS

# PREFACE

If there are two available canonical representations of the elements of a set it is often possible to establish relations between them. This applies in particular to polynomial rings over fields, as the elements in them admit both additive and multiplicative representations. We mean the coefficient representation and the prime factorization of polynomials, respectively. Interesting questions arise from looking for polynomials both of whose representations are specialized in some way or other. We are led in this way to the problem of finding those polynomials over a field that are lacunary and fully reducible by which we mean, respectively, the disappearance of certain coefficients and the splitting into linear factors (over the fundamental field).

More than 20 years ago I recognized that the extension of the factorization theory of finite abelian groups, developed by Hajós [2], was made possible by the determination of certain fully reducible lacunary polynomials over finite prime fields. Moreover, a whole series of new applications in different domains of Algebra and the Theory of Numbers have been derived from this same source. As, so far, the use of lacunary polynomials has been indispensable for the derivation of these applications, a large measure of independent significance must be accorded to them.

Even at that time I considered the generalization of my investigations to arbitrary finite fields instead of prime fields; I was frightened to do so, however, faced with the enormous difficulties involved. Still, as the number of applications of my old results increased, three years ago I took up again the lost thread. The difficulties turned out to be even greater than I had expected. The battle was hard-fought, but successful. In the meantime the material has become so voluminous that I had to decide to publish both my old and my new results in book form. The realization of this intention is the present volume.

It is only reasonable to count in with a given theory parts of other domains closely connected with it. We may therefore say that a second branch of Hajós's theory is developed in the present book. This theory is in a state of continuous development and it is indeed the most beautiful of its features that it displays a surprising number of connections with remote branches of Mathematics which appear in the form of auxiliary investigations, equivalent reformulations and applications (cf. Rédei [10]).

The most important of these are the equivalence of Hajós's main theorem and the famous conjecture of Minkowski on the marginal case of a homogeneous system of linear diophantine inequalities. The latter marginal case, solved in this way, was the start of Hajós's theory. It is the problem of a certain marginal case, here too, that will give us most trouble. We may therefore say that the appearance of difficult marginal cases is a recurrent phenomenon of Hajós's theory.

The book is divided into six chapters, the first five dealing mainly with three problems on fully reducible lacunary polynomials over finite fields; these will be called Problems I, II and III. Problem I is much easier than the others and of less importance for the applications. I consider Problem II to be the main problem of the book. Problem III is an auxiliary problem of Problem II, of interest also in its own right, to which the "marginal case" of the latter is reducible. Several further auxiliary problems connected with polynomials and power series over finite fields will be solved, these likewise being of independent significance. The applications are given in the last chapter. (To give the reader an idea of the diversity of these applications, he is referred to the headings of §§ 34–39.)

I have done my utmost to make the argument enjoyable for the reader. For this purpose I have inserted many explanatory notes in the text which may also be useful for the more experienced reader.

At the end of the book there are a number of unsolved problems. Interested readers will, however, surely find even more problems. I wish them the best of luck!

I have dedicated my book to Professors G. Hajós, L. Kalmár and P. Turán out of gratitude, as they assisted me in scientific problems on many occasions and for a long time, also in connection with the material of this book.

I should also like to express my thanks to my younger friends for the interest they have shown in my research. Among them I must mention the names of B. Csákány, L. Gehér, S. Lajos, G. Pollák, O. Steinfeld, E. Wittmann and especially L. Megyesi, many of whose valuable suggestions have been used in my book.

*The Author*

CHAPTER I

# PRELIMINARIES AND FORMULATION OF PROBLEMS I, II, III

Of these preliminary considerations § 3 is also of independent interest.

## § 1. Some basic concepts and notations

Conventional concepts and notations will often be used without explanation or with only a brief hint.

We shall write the *cardinal number* of a set as $O(\mathsf{S})$ (thus also the *order* of an algebraic structure). A set with cardinal number 1 will be identified with its element.

If $f(x)$ is a polynomial or a power series (terminating on the left or on the right), then the set of exponents occurring in the (non-vanishing) terms of $f(x)$ is called the *exponent set* of $f(x)$, denoted by $\mathfrak{E}(f)=\mathfrak{E}(f(x))$. The polynomial $f(x)$ is said to be *lacunary* if there are elements $i, j \in \mathfrak{E}(f)$ for which $i+2 \leqq j$ and not all the $i+1, \ldots, j-1$ lie in $\mathfrak{E}(f)$; in order to avoid exceptional cases, $f(x)$ is also called lacunary if it is equal to 0 or a monomial. If

$$i, j \in \mathfrak{E}(f); \qquad i+2 \leqq j; \quad i+1, \ldots, j-1 \notin \mathfrak{E}(f),$$

we say that $f(x)$ has a *gap* between $i$ and $j$ (or from $i$ up to $j$) and we shall call $j-i$ the *length* of this gap.

Let $\mathsf{F}$ denote a finite field.

For the *characteristic*, the *degree* and the *order* of $\mathsf{F}$ we introduce the notations

$$p=\mathfrak{ch}(\mathsf{F}), \quad n=\mathsf{F}^{\circ}, \quad q=O(\mathsf{F})=p^n.$$

(Thus $p$ may denote any prime number and $n$ any natural number). Already here it is necessary to point out that in part of what follows the case $p=2$ will prove to be trivial and will therefore be disregarded where appropriate.

$\mathscr{Z}$ and $\mathscr{N}$ denote the *ring of integers* and the *semiring of non-negative integers*, respectively. We shall not in every possible case use these notations; e.g., for $i \in \mathcal{N}$, we shall often write $i=0, 1, \ldots$,

$\left(\frac{a}{p}\right)$ is the Legendre symbol ($a \in \mathscr{Z}$).

"$|$" is the usual symbol for divisibility.

For $k \in \mathcal{N}$ and for elements $\alpha$, $\beta$ of an integral domain, $\alpha^k \| \beta$ signifies the validity of $\alpha^k | \beta$ and $\alpha^{k+1} \nmid \beta$, provided that $\alpha$ and $\beta$ are different from 0, and that $\alpha$ is not a unit.

For $a, b \in \mathcal{N} \setminus 0$, $a \underset{1}{|} b$ denotes the validity of $a|b$ and $a-1|b-1$. The transitivity of this relation is obvious. It is clear that if $b=p^k$ ($k=1, 2, \ldots$) $a \underset{1}{|} b$ holds if and only if $a=p^d$ ($d|k$).

$P$ denotes the order of a subfield of $\mathsf{F}$, while $\mathsf{F}_P$ denotes this subfield itself. The numbers $P$ are the powers $p^k$ ($k|n$), where $k=\mathsf{F}_p^\circ$. They can obviously also be characterized by $P \underset{1}{|} q$. Clearly $\mathsf{F}_q=\mathsf{F}$ holds and $\mathsf{F}_p$ is the prime field of $\mathsf{F}$. Occasionally in this book we shall denote arbitrary divisors of $q$ by $P$, but we shall draw the reader's attention to this whenever we do so.

As usual, the integers and, more generally, the rational numbers with a denominator prime to $p$, which we shall call rational *p-integers*, will also denote the elements of $\mathsf{F}_p$ (the unity element of $\mathsf{F}_p$ may therefore be denoted by 1). Accordingly, $\mathsf{F}_p$ is said to consist of the rational elements of $\mathsf{F}$, which will not lead to any misunderstanding. (In other structures with a unity element the latter will often also be denoted by 1.)

Since the mapping

$$\xi \to \xi^P \quad (\xi \in \mathsf{F})$$

is a *generating automorphism* for the *field extension* $\mathsf{F}|\mathsf{F}_P$, the corresponding *relative norm* of $\xi$ is given by

$$N_{\mathsf{F}|\mathsf{F}_P}(\xi) = \xi^{\frac{q-1}{P-1}} \left(\xi^{1+P+P^2+\cdots+\frac{q}{P}}\right).$$

In particular, $\xi \to \xi^p$ is a *generating* (*absolute*) *automorphism* of $\mathsf{F}$; moreover

$$N_{\mathsf{F}}(\xi) = \xi^{\frac{q-1}{p-1}}$$

is the (*absolute*) *norm* of $\xi$.

$\bar{\mathsf{F}}$ denotes the *algebraic closure* of $\mathsf{F}$ (consequently also of $\mathsf{F}_p$),

For all $\alpha \in \bar{\mathsf{F}}$ and for every rational number $r$ we define the *power* $\alpha^r$, which is in general *multiple-valued*, as follows: we write $r$ in its canonical form

$$r = \frac{i}{k} \qquad (i \in \mathscr{Z},\ k \in \mathcal{N} \setminus 0;\ (i, k) = 1)$$

and understand by $\alpha^r$ any solution of the equation

$$x^k = \alpha^i,$$

lying in $\bar{\mathsf{F}}$. Accordingly, $\alpha^r$ may also be called (in the case $k \geqq 2$) a *k-th root* of $\alpha^i$ (for which the notation $\sqrt[k]{\alpha^i}$ is usually employed; however, we shall make little use of the latter symbol). Besides, only that case will generally occur in which $r = \frac{1}{k}$ $(k = 2, 3, \ldots)$, $\alpha \in \mathsf{F}$ and, furthermore, all values of $\alpha^r$ are contained in $\mathsf{F}$.

$\mathsf{F}[x]$ is the *polynomial ring* over $\mathsf{F}$ (with the indeterminate $x$).

$\mathsf{F}(x)$ is the *quotient field* of $\mathsf{F}[x]$. Its elements are called polynomial quotients (over $\mathsf{F}$). Though it is customary to call the latter "rational functions" and $\mathsf{F}(x)$ itself a "rational function field", we shall not here use these incorrect names which may easily lead to misunderstandings, especially as we shall have to deal also with a *function* $\xi \to f(\xi)$ $(\xi \in \mathsf{F})$, *assigned* to a polynomial $f(x) \in \mathsf{F}(x)$.

In $\mathsf{F}(x)$, apart from addition and multiplication, we shall treat *iteration*, too, as a (non-commutative) composition by which the element $g(f(x))$ is assigned to a pair $f(x), g(x) \in \mathsf{F}(x)$. Iteration is defined similarly in $\mathsf{F}[x]$, since the *iterated polynomial* $g(f(x))$ lies in $\mathsf{F}[x]$, whenever $f(x), g(x) \in \mathsf{F}[x]$. For a fixed $f(x) \in \mathsf{F}[x]$ the set of all the iterated polynomials $g(f(x))(g(x) \in \mathsf{F}[x])$ constitute a subring of $\mathsf{F}[x]$, denoted by $\mathsf{F}[f(x)]$. We shall, however, make little use of this notation.

For $f(x) \in \mathsf{F}[x]$ the greatest natural number $k$ with $f(x) \in \mathsf{F}[x^k]$ is said to be the *exponent* of $f(x)$.

In particular, $\mathsf{F}[x^p]$ is the subring of $\mathsf{F}[x]$ consisting of all the $f(x) \in \mathsf{F}[x]$ with *vanishing derivative* $f'(x)$ which we shall occasionally call *degenerate polynomials*. It should be noted that

$$\mathsf{F}[x^p] = \mathsf{F}[x]^p$$

(where on the right-hand side we raise to the *p-th* power element by element).

In the case $k = 0$

$$g_1(\ldots(g_k(f(x)))\ldots)$$

means $f(x)$, of course. In particular, $g_1(\ldots(g_k(x))\ldots) = x$, if $k = 0$.

By $\mathsf{F}((x))$ and $\mathsf{F}\left(\left(\frac{1}{x}\right)\right)$ we denote the completions of $\mathsf{F}(x)$, belonging to the places 0 and $\infty$, respectively. (See Hasse [2]). Accordingly, $\mathsf{F}((x))$ is the *field of the power series*

$$\sum_{-\infty \ll i < \infty} \alpha_i x^i, \qquad (\alpha_i \in \mathsf{F})$$

*terminating on the left*, and $\mathsf{F}\left(\left(\frac{1}{x}\right)\right)$ is the *field of the power series*

$$\sum_{-\infty<i\ll\infty} \alpha_i x^i, \qquad (\alpha_i \in \mathsf{F})$$

*terminating on the right*.

We denote by $\mathsf{F}[[x]]$ the ring of the *integer* elements

$$\sum_{0\leqq i<\infty} \alpha_i x^i$$

of $\mathsf{F}((x))$. Thus $\mathsf{F}[x]$ is a subring of $\mathsf{F}[[x]]$. It should be noted that the integer elements of $\mathsf{F}(x)$, i.e. the elements of $\mathsf{F}[x]$, are integers also in $\mathsf{F}((x))$. (It should similarly be noted that $\mathsf{F}(x)$ is contained in $\mathsf{F}((x))$ as well as in $\mathsf{F}\left(\left(\frac{1}{x}\right)\right)$ as a subfield and therefore its elements can in general be developed into a power series in two ways.)

As usual, we define the *degree* of a polynomial $f(x)$ or, more generally, of a power series $f(x)$, terminating on the right, as the maximum element of the exponent set $\mathfrak{E}(f)$, denoting it by $f^\circ$. In the case $f(x)=0$ this definition fails; however, we shall then write $f^\circ=-\infty$. If $f(x)\neq 0$, we define the *second degree* $f^{\circ\circ}$ of $f(x)$ as the (first) degree of the part of $f(x)$ remaining after the term of degree $f^\circ$ has been cancelled. The difference

$$f^\downarrow = f^\circ - f^{\circ\circ} (\geqq 1)$$

will be called the *sinking* of $f(x)$. If $f^\downarrow \geqq 2$ and $f^{\circ\circ}\geqq 0$, then $f(x)$ has a gap of length $f^\downarrow$ between $f^{\circ\circ}$ and $f^\circ$. (Where necessary, we shall write $f(x)^\circ$, $f(x)^{\circ\circ}$ and $f(x)^\downarrow$ for $f^\circ$, $f^{\circ\circ}$ and $f^\downarrow$, respectively). It should be emphasized that the $f(x)$ with a sinking $\geqq 2$ are also lacunary.

Since $\mathsf{F}(x)\subset\mathsf{F}\left(\left(\frac{1}{x}\right)\right)$ the concepts of degree, second degree and sinking are to be considered as also defined for the polynomial quotients

$$u(x)=\frac{f(x)}{g(x)} \qquad (f(x),\ g(x)\in\mathsf{F}[x];\ g(x)\neq 0).$$

Clearly, $u^\circ = f^\circ - g^\circ$ holds. (Of course, another concept of degree is sometimes introduced in $\mathsf{F}(x)$ for certain purposes, with which we shall, however, not have to deal in what follows.)

The coefficients $k_0, k_1, \ldots$ occurring in the "canonic" *P-adic representation*

$$k = k_0 + k_1 P + \cdots \qquad (k_i = 0, \ldots, P-1;\ i = 0, 1, \ldots)$$

of the $p$-integer rational numbers are called the *P-adic numerals* of $k$. Briefly

$k_i$ is said to be the $i$-th numeral. The maximum $\max k_i$ of the numerals is called the *P-adic height* of $k$, denoted by $k^{\circ_P}$. We define the $P$-adic height of a non-empty set of $p$-integers as the maximum of the $P$-adic heights of its elements.

By the *P-degree* $f^{\circ_P}(=f(x)^{\circ_P})$ of a polynomial (or of a power series) $f(x)$ we shall understand the $P$-adic height of the exponent set $\mathfrak{E}(f)$; we shall write $f^{\circ_P}=0$ if $f(x)=0$. (The similarity between the notation for the $P$-adic height of a number and that for the $P$-degree of a polynomial will give no trouble.)

It should be noted that a polynomial $f(x)$ with $f^\circ \leqq 1$ is called *linear*; this means that a constant $f(x)$ is also regarded as linear. (The transformation $x \to f(x)$, however, is said to be linear only if $f^\circ = 1$.) The polynomial $f(x)$ is called *P-linear* if $f^{\circ_P} \leqq 1$. Thus the $P$-linear $f(x) \in \mathsf{F}[x]$ are those of the form

$$f(x) = \alpha_0 + \alpha_1 x + \alpha_P x^P + \alpha_{P+1} x^{P+1} + \alpha_{P^2} x^{P^2} + \dots, \qquad (\alpha_i \in \mathsf{F})$$

so they are in general (strongly) lacunary. It is trivial that for a non-constant $P$-linear polynomial $f(x)$ there always exists a unique $l=0, 1, \dots$ such that

$$P^l \leqq f^{\circ_P} \leqq \frac{P^{l+1}-1}{P-1} \;(= P^l + P^{l-1} + \dots + P + 1).$$

A polynomial $f(x) \in \mathsf{F}[x]$ is called *fully reducible* if it splits into a product of linear factors in $\mathsf{F}[x]$, i.e. its splitting field over $\mathsf{F}$ is equal to $\mathsf{F}$. The constant $f(x) \neq 0$ are regarded as fully reducible polynomials.

As is well known, the fully reducible polynomial $x^q - x$ plays a very important role in $\mathsf{F}[x]$; we call it the *Galois binomial*. It follows from

$$x^q - x = \prod_{\xi \in \mathsf{F}} (x - \xi)$$

that a polynomial $f(x) \in \mathsf{F}[x]$ is fully reducible if and only if it is a divisor of a power of $x^q - x$. The known divisors

$$x^k - \alpha \qquad (k \mid (q-1),\ \alpha \in \mathsf{F}^k \setminus 0)$$

of $x^q - x$ are called the *Euler binomials*. If $k \mid q-1$, there are exactly $\dfrac{q-1}{k}$ Euler binomials of degree $k$ in $\mathsf{F}[x]$ and their product is equal to $x^{q-1}-1$ (that is the only Euler binomial of degree $q-1$). The linear Euler binomials are all the

$$x - \alpha \qquad (\alpha \in \mathsf{F} \setminus 0).$$

For a polynomial $f(x) \in \mathsf{F}[x]$ we shall call

$$F(x) = \prod_{\varrho \in \bar{\mathsf{F}},\, f(\varrho)=0} (x - \varrho)$$

the *zero polynomial* of $f(x)$. It is evident that $F(x) \mid x^q - x$ holds if and only if $f(x)$ is fully reducible.

A polynomial $f(x) \in \mathsf{F}[x]$ is called an $(\mathsf{F}, \mathsf{F}_P)$-*polynomial* if all the values of $f(\xi)$ $(\xi \in \mathsf{F})$ are contained in a given subfield $\mathsf{F}_P$ of $\mathsf{F}$. In other words, this means that $\xi \to f(\xi)$ is a mapping of $\mathsf{F}$ into $\mathsf{F}_P$. In particular, the $(\mathsf{F}, \mathsf{F}_P)$-polynomials will also be called *rational-valued polynomials*. The case $P=q$ is uninteresting, as every $f(x) \in \mathsf{F}[x]$ is an $(\mathsf{F}, \mathsf{F})$-polynomial; we do not, however, wish to exclude this case. Less precisely, the $(\mathsf{F}, \mathsf{F}_P)$-polynomials will sometimes also be called *polynomials* (from $\mathsf{F}[x]$) *with a restricted range* (which is not correct if $P=q$). The $(\mathsf{F}, \mathsf{F}_P)$-polynomials will be thoroughly investigated in § 3; it should, however, be pointed out already here that the existence of non-constant $(\mathsf{F}, \mathsf{F}_P)$-polynomials follows from the well-known fact that in $\mathsf{F}$ all functions are already represented by the polynomials of $\mathsf{F}[x]$. We must also emphasize that the condition $f(x) \in \mathsf{F}_P[x]$ is neither necessary nor sufficient for an $f(x) \in \mathsf{F}[x]$ to be an $(\mathsf{F}, \mathsf{F}_P)$-polynomial; later (in § 3) we shall gain much more exact information on this point.

In any case when in a (commutative) Euclidean ring $\mathsf{R}$ a unique *Euclidean division*

$$\alpha = \beta\varkappa + \varrho \qquad (\alpha, \beta, \varkappa, \varrho \in \mathsf{R};\ \beta \neq 0)$$

is defined according to which for each pair of elements $\alpha$, $\beta$ the *(Euclidean) quotient* and *residue* $\varkappa$ and $\varrho$ are uniquely determined, we introduce the notation

$$\varkappa = \left[\frac{\alpha}{\beta}\right], \qquad \varrho = (\alpha)_\beta$$

$\left(\frac{\alpha}{\beta}\right.$ must be thought of as an element of the quotient field of $\left.\mathsf{R}\right)$. They will then be called the *integer part* of $\frac{\alpha}{\beta}$ and the *residue* of $\alpha$ mod $\beta$, respectively. Thus, corresponding to the classical cases $\mathsf{R}=\mathscr{Z}$, $\mathsf{F}[x]$, the symbols

$$\left[\frac{a}{b}\right], (a)_b \ (a, b \in \mathscr{Z}; b \neq 0); \quad \left[\frac{f(x)}{g(x)}\right], (f(x))_{g(x)} \ (f(x), g(x) \in \mathsf{F}[x]; g(x) \neq 0)$$

are meaningful (the uniqueness of which being assured by the usual prescriptions $0 \leqq (a)_b < b$ and $((f(x))_{g(x)})^\circ < g^\circ$).

NOTE 1. Some of our definitions and simple proofs, depending on $\mathsf{F}$, retain their validity for other fields (instead of $\mathsf{F}$), a fact of which we shall avail ourselves (without reference) from time to time in this book. Accordingly, for example, $\bar{\mathsf{F}}(x)$ and $\bar{\mathsf{F}}((x))$ can be considered as defined.

NOTE 2. Some concepts connected with polynomials will be carried over to power series without further reference.

NOTE 3. Although occasionally $P$ could signify any power $p, p^2, \ldots$ or even any natural number $\geqq 2$, we shall be interested constantly only in the case $P\underset{1}{|}q$ (or exceptionally in the case $P|q$).

NOTE 4. We shall always endeavour to work not with the single coefficients or terms of a polynomial $f(x)$ but rather with its different additive decompositions of the form

$$f(x) = \sum_{i=1}^{k} x^{e_i} f_i(x)$$

and with the "constituents" $f_i(x)$ occurring in the latter. Our considerations will gain considerably in elegance from this procedure.

NOTE 5. If two problems $A$ and $B$ are connected in such a way that any solution of $A$ is necessarily a solution of $B$, then $B$ may be considered an "incomplete reduction" of $A$, because in such cases, after the determination of the solutions of $B$, we have to select from them those that are also solutions of $A$, in order to obtain all the solutions of $A$. We shall repeatedly have to deal with such (incomplete) reductions, without calling the reader's attention to this. The selection of suitable solutions will always be rather difficult.

NOTE 6. In particular, we have an incomplete reduction when, by the "principle of the extension of the domain of solutions", we try to obtain the solutions of a system of equations that belong to a certain structure in such a way that first we determine all solutions lying in a suitably extended structure.

Let us end this chapter with

NOTE 7. Because of the above convention the reader should always understand the $p$-integers used in our definitions and statements as being elements of $\mathsf{F}_p$, even if we give no more than a slight indication to this effect; this may often be necessary in order to understand what we have said.

## § 2. Invariants of polynomials

In $\mathsf{F}[x]$ and in some of its subsets we shall have to deal with several kinds of invariants about which we wish to summarize here briefly the knowledge necessary for our purpose.

First of all let $\mathsf{S}$ denote an arbitrary set and $\mathscr{R}$ a binary relation in it. A property $\mathscr{P}$ (which the elements of $\mathsf{S}$ may possess) is said to be *$\mathscr{R}$-invariant* (or to be an *$\mathscr{R}$-invariant*) if for every element $a \in \mathsf{S}$, possessing the property $\mathscr{P}$, all elements $b \in \mathsf{S}$ with $a\mathscr{R}b$ also possess the property $\mathscr{P}$. (Following a suggestion by

B. Csákány, we shall call such invariants *relation invariants*.) We discuss two special cases.

First let $\mathcal{R}$ be an equivalence in S. The corresponding equivalence classes will be called *$\mathcal{R}$-classes* and elements lying in the same class are said to be *$\mathcal{R}$-equivalent*. In this case we shall also say of the $\mathcal{R}$-invariants that they are *class invariants*.

Secondly, let us assume that a set $\mathfrak{M}$ of mappings of S into itself is given. We define a relation $\mathcal{R}=\mathcal{R}_{\mathfrak{M}}$ in such a way that for $a, b \in \mathsf{S}$ the relation $a\mathcal{R}b$ holds if and only if $a$ is carried over into $b$ by a mapping from $\mathfrak{M}$. Instead of an $\mathcal{R}$-invariant we shall speak in this case of an $\mathfrak{M}$-invariant calling it a *mapping invariant*, too.

An important (common) special case arises when $\mathfrak{M}=\mathfrak{P}$ is a permutation group (of S). This defines an equivalence in S, a class being formed by the elements transitively connected with one another under $\mathfrak{P}$; for this reason we may speak in the corresponding sense of $\mathfrak{P}$-classes, $\mathfrak{P}$-equivalence and $\mathfrak{P}$-equivalent elements. In this case the $\mathfrak{P}$-invariants can be thought of both as mapping- and as class invariants.

We must also note the following. Until now we have been dealing with invariant properties but it is also possible to speak, for example, of invariant functions and invariant subsets in a similar sense. A function $\Phi$ with the definition domain S is naturally called $\mathcal{R}$-invariant if the value $\Phi(a)$ of $\Phi$ for $a \in \mathsf{S}$ is $\mathcal{R}$-invariant, i.e. if $\Phi(a)=\Phi(b)$ always holds for $a\mathcal{R}b$. A subset $\mathfrak{A}$ of S is called $\mathcal{R}$-invariant if the property of an element $a \in \mathsf{S}$ to belong to $\mathfrak{A}$ is $\mathcal{R}$-invariant, i.e. $a \in \mathfrak{A}$ and $a\mathcal{R}b$ always imply $b \in \mathfrak{A}$. If $\mathcal{R}$ is here an equivalence, then the $\mathcal{R}$-invariant subsets of S coincide with the unions of $\mathcal{R}$-classes.

We now proceed to discuss the concepts of invariance to be applied in F[$x$]. Some of these concepts are based on the *(inner) linear transformations*

$$f(x)\to f(\alpha x+\beta) \qquad \big(f(x)\in \mathsf{F}[x];\ \alpha, \beta\in \mathsf{F};\ \alpha\neq 0\big) \tag{1}$$

and on the *outer linear transformations*

$$f(x)\to \gamma f(x)+\delta \qquad (\gamma, \delta\in \mathsf{F};\ \gamma\neq 0), \tag{2}$$

furthermore on the *linear double transformations*

$$f(x)\to \gamma f(\alpha x+\beta)+\delta \tag{3}$$

(composed of (1) and (2)). The four special cases

$$f(x)\to f(\alpha x), \tag{4}$$

$$f(x)\to f(x+\beta), \tag{5}$$

$$f(x)\to \gamma f(x), \tag{6}$$

$$f(x)\to f(x)+\delta \tag{7}$$

are called *rotations*, *translations*, *similarity transformations* and *parallel transformations*, respectively. The transformations of each of the different kinds enumerated under (1)—(7) form a permutation group, denoted by $\mathfrak{P}_i$ ($i=1, \ldots, 7$). (All these groups are subgroups of $\mathfrak{P}_3$ and $\mathfrak{P}_3$ is the direct product of $\mathfrak{P}_1$ and $\mathfrak{P}_2$. Moreover, the relations $\mathfrak{P}_1=\mathfrak{P}_4\mathfrak{P}_5$ and $\mathfrak{P}_2=\mathfrak{P}_6\mathfrak{P}_7$ hold and, finally, $\mathfrak{P}_5$ and $\mathfrak{P}_7$ are normal in $\mathfrak{P}_3$.) In connection with $(\mathsf{S}=)\mathsf{F}[x]$ and $(\mathfrak{P}=)\mathfrak{P}_i$ ($i=1, 3, 4, \ldots, 7$) we shall avail ourselves of the concepts introduced above. (We do not care about the case $i=2$ with $\gamma \neq 1$.) Accordingly we shall use the names summarized in the following table:

| $\mathfrak{P}$ | $\mathfrak{P}$-classes | $\mathfrak{P}$-equivalent | $\mathfrak{P}$-invariant |
|---|---|---|---|
| $\mathfrak{P}_1$ | classes | equivalent | invariant |
| $\mathfrak{P}_3$ | full classes | equivalent in the wider sense | fully invariant |
| $\mathfrak{P}_4$ | rotation classes | rotation equivalent | rotation invariant |
| $\mathfrak{P}_5$ | translation classes | translation equivalent | translation invariant |
| $\mathfrak{P}_6$ | similarity classes | associated | similarity invariant |
| $\mathfrak{P}_7$ | parallel classes | parallel | parallel invariant |

(The short names for the case $\mathfrak{P}_1$ emphasize the importance and frequent occurrence of this case.)

The case $(\mathfrak{P}=)\mathfrak{P}_6$ possesses the particular significance that in almost all questions referring to $\mathsf{F}[x]$ (divisibibity, full reducibility, lacunarity, degree, second degree, sinking of the elements of $\mathsf{F}[x]$ etc.), associated polynomials are completely equivalent with one another. We therefore mark out a representative system of the similarity classes in $\mathsf{F}[x]$. The most convenient way of doing this is to choose, apart from 0, the monic polynomials $f(x)$, i.e., those with initial coefficient 1, as the representative system of the similarity classes. Because of this we shall often restrict our considerations to monic polynomials (without, however, any loss of generality).

It is inconvenient that the property of being monic is not invariant. It is therefore best to agree to call two monic polynomials from $\mathsf{F}[x]$ *equivalent* or *rotation equivalent* if this is the case apart from a norming factor. This means that the monic polynomials, equivalent or rotation equivalent to a monic $f(x) \in \mathsf{F}[x]$ are given by

$$\alpha^{-f^\circ} f(\alpha x+\beta) \quad \text{or, respectively,} \quad \alpha^{-f^\circ} f(\alpha x) \qquad (\alpha, \beta \in \mathsf{F};\ \alpha \neq 0).$$

(It should be noted that, since the set of the monic polynomials from $\mathsf{F}[x]$ are translation invariant, the translation equivalence of monic polynomials can be interpreted in the ordinary way.)

Henceforth let $f(x) \underset{1}{\in} \mathsf{F}[x]$ indicate that $f(x)$ is a monic polynomial from $\mathsf{F}[x]$.

As an important example for invariants, we note that the degree $f^\circ$ of $f(x) \in \mathsf{F}[x]$ is invariant (and in the case $f^\circ \geqq 1$ even fully invariant). We prove

**Proposition 1.** *If the degree $f^\circ$ of a polynomial $f(x) \in \mathsf{F}[x]$ is a power $>1$ of $p$ and its second degree $f^{\circ\circ}$ is positive, then the latter and, consequently, also the sinking $f^\downarrow$ are invariant.*

On account of the supposition we have

$$f(x) = \varrho x^{p^k} + g(x) \qquad (\varrho \in \mathsf{F}\setminus 0;\ k \geqq 1;\ g(x) \in \mathsf{F}[x];\ 1 \leqq g^\circ = f^{\circ\circ} < p^k)$$

with suitable $\varrho$, $k$, $g(x)$. Thus

$$f(\alpha x + \beta) = \varrho \alpha^{p^k} x^{p^k} + g(\alpha x + \beta) + \varrho \beta^{p^k} \qquad (\alpha, \beta \in \mathsf{F}),$$

whence the validity of Proposition 1 follows.

If $p \nmid f^\circ$ the second degree $f^{\circ\circ}$ is not even translation invariant. Moreover, the following proposition is familiar:

**Proposition 2.** *For a polynomial $f(x) \in \mathsf{F}[x]$ with $p \nmid f^\circ$ there exists a unique translation $x \to x + \xi$ $(\xi \in \mathsf{F})$ with $f(x+\xi)^\downarrow \geqq 2$.*

We prove

**Proposition 3.** *The P-degree $f^{\circ_P}$ of $f(x) \in \mathsf{F}[x]$ is invariant. In particular, this implies that the P-linearity is an invariant property* (*in* $\mathsf{F}[x]$).

The rotation invariance of $f^{\circ_P}$ is trivial; it is therefore sufficient to prove the translation invariance. It suffices for this purpose to show

$$(f(x+\xi))^{\circ_P} \leqq f^{\circ_P}.$$

Now the $f(x)$ with $f^{\circ_P} \leqq k$ are obviously obtained for a fixed $k = 0, \ldots, P-1$ by starting from polynomials of the form

$$(x+1)^k (x^P+1)^k (x^{P^2}+1)^k \ldots (x^{P^l}+1)^k \qquad (l = 0, 1, \ldots)$$

(after multiplying out) giving to the terms arbitrary coefficients from $\mathsf{F}$. Our assertion is easily verified, from this observation and from the fact that

$$(x+\xi)^{P^i} + 1 = x^{P^i} + (\xi^{P^i} + 1) \qquad (i = 0, 1, \ldots).$$

Thus we have proved Proposition 3.

We notice that the full reducibility of polynomials in $\mathsf{F}[x]$ is obviously invariant and similarity invariant but, in general, it is not parallel invariant and consequently not fully invariant. (Exceptionally, full reducibility is fully invariant for the linear polynomials of $\mathsf{F}[x]\setminus 0$.)

On the subject of the occurrence of parallel polynomials in connection with iterated polynomials we wish to note the following. For a fully reducible polynomial $f(x) \in \mathsf{F}[x]$

$$f(x) = \prod_{i=1}^{k} (x - \varrho_i) \qquad (\varrho_i \in \mathsf{F})$$

holds with suitable $\varrho_1, \ldots, \varrho_k$, whence it follows that for a $g(x) \in \mathsf{F}[x]$ the iterated polynomial $f(g(x))$ is equal to the product of the parallel polynomials

$$g(x) - \varrho_1, \ldots, g(x) - \varrho_k .$$

We now consider all polynomials

$$f(x) = \alpha_0 + \alpha_1 x \cdots + \alpha_k x^k \qquad (\alpha_0, \ldots, \alpha_k \in \mathsf{F}) \tag{8}$$

for a fixed $k = 0, 1, \ldots$ . We say that these polynomials are of the *formal degree* $k$. (That means $f^{\circ} \leqq k$.) It is obvious that the set of these polynomials undergoes a permutation by

$$f(x) \to x^k f\left(\frac{1}{x}\right). \tag{9}$$

This permutation is called the *reciprocal transformation* (or *reciprocity*) of the set of polynomials of formal degree $k$ from $\mathsf{F}[x]$. Since it is an involution (i.e. of order 2), it forms, together with the identity mapping, a permutation group of order 2 of this last set. The resulting classes are called *reciprocity classes* and two polynomials belonging to the same class are said to be *reciprocal* (to one another). (This means that the left- and right-hand sides of (9) are reciprocal polynomials which may also be equal.) The corresponding invariant properties will be called *reciprocity invariant*. It should be noticed that the concepts introduced here are dependent on $k$. In the applications the $k$, chosen in advance, will always be indicated in some way.

Finally, invariants will also occur in $\mathsf{F}[x]$ in connection with the mappings

$$f(x) \to f(x)^k \qquad (k = 2, 3, \ldots)$$

and

$$f(x) \to f'(x).$$

Concerning these the following trivial statements hold:

PROPOSITION 4. *The sinking $f^{\downarrow}$ of polynomials $f(x) \in \mathsf{F}[x] \setminus 0$ is invariant with respect to the operation of raising to a power prime to $p$, i.e.*

$$(f(x)^k)^{\downarrow} = f^{\downarrow} \qquad (p \nmid k;\ k \in \mathcal{N})$$

*holds.*

PROPOSITION 5. *The sinking of polynomials $f(x) \in \mathsf{F}[x]$ of degree and second degree prime to p is invariant with respect to differentiation, i.e.*

$$f'(x)^{\downarrow} = f^{\downarrow} \qquad (p \nmid f^{\circ}, f^{\circ\circ})$$

*holds.*

It is not our purpose to record all invariance properties occurring in $\mathsf{F}[x]$. We shall cite some examples here; other cases will be discussed later on, if necessary.

EXAMPLE 1. $\mathsf{F}[x^p]$ is a fully invariant subring of $\mathsf{F}[x]$; this is a consequence of

$$\gamma f\big((\alpha x + \beta)^p\big) + \delta = \gamma f(\alpha^p x^p + \beta^p) + \delta \qquad (f(x) \in \mathsf{F}[x];\ \alpha, \beta, \gamma, \delta \in \mathsf{F}).$$

EXAMPLE 2. For every $P$ all $(\mathsf{F}, \mathsf{F}_P)$-polynomials, by their definition, constitute an invariant subring of $\mathsf{F}[x]$.

## § 3. Polynomials with a restricted range

For an arbitrary subfield $\mathsf{F}_P$ of $\mathsf{F}$ we shall need an exact knowledge of the $P$-linear $(\mathsf{F}, \mathsf{F}_P)$-polynomials. In order not to be delayed by it later, we wish to settle this question now. We actually determine all the $(\mathsf{F}, \mathsf{F}_P)$-polynomials (not only the $P$-linear ones), since this greater generality does not cause additional work but rather tends to simplify matters. The determination will be carried out in three ways, all of which will later be taken into consideration.

First of all we prove

THEOREM 1. *A polynomial $f(x) \in \mathsf{F}[x]$ is an $(\mathsf{F}, \mathsf{F}_P)$-polynomial if and only if*

$$x^q - x \mid f(x)^P - f(x) \tag{1}$$

*holds.*

We call (1) the *divisibility condition* for $(\mathsf{F}, \mathsf{F}_P)$-polynomials.

In order to prove this theorem, it should be remembered that a $\xi \in \mathsf{F}$ lies in $\mathsf{F}_P$ if and only if $\xi^P = \xi$. Consequently $f(x)$ is an $(\mathsf{F}, \mathsf{F}_P)$-polynomial if and only if it is always true that

$$f(\eta)^P = f(\eta) \qquad (\eta \in \mathsf{F}),$$

i.e. the congruence

$$f(x)^P \equiv f(x) \qquad (\mathrm{mod}\ x^q - x)$$

holds. This proves Theorem 1.

Since the que stionof whether an $f(x)$ is an $(\mathsf{F}, \mathsf{F}_P)$-polynomial or not, depends only on the residue class $f(x)$ (mod $x^q - x$), we shall henceforth restrict our considerations to the $(\mathsf{F}, \mathsf{F}_P)$-polynomials of degree $\leqq q-1$. We prove

THEOREM 2. *A polynomial*

$$f(x) = \sum_{i=0}^{q-1} \alpha_i x^i \qquad (\alpha_i \in \mathsf{F}) \tag{2}$$

*is an* $(\mathsf{F}, \mathsf{F}_P)$*-polynomial if and only if its coefficients satisfy the conditions*

$$\alpha_j = \alpha_i^P \quad (j \equiv Pi \,(\mathrm{mod}\, q-1);\ i, j = 0, \ldots, q-2); \quad \alpha_{q-1} = \alpha_{q-1}^P. \tag{3}$$

We call (3) the *conjugacy conditions* for the (coefficients of the) $(\mathsf{F}, \mathsf{F}_P)$-polynomials of degree $\leqq q-1$.

It suffices to prove that for (2) condition (1) is equivalent to (3). Because of (2) the condition (1) can be written in the form

$$\sum_{i=0}^{q-1} \alpha_i x^i \equiv \sum_{i=0}^{q-1} \alpha_i^P x^{Pi} \,(\mathrm{mod}\, x^q - x). \tag{4}$$

On the other hand, we obviously have

$$x^{Pi} \equiv \begin{cases} x^{(Pi)_{q-1}} (\mathrm{mod}\, x^q - x) & (i = 0, \ldots, q-2), \\ x^{q-1} \quad (\mathrm{mod}\, x^q - x) & (i = q-1). \end{cases}$$

Thus, after the separation of the terms belonging to $i = q-1$, (4) becomes

$$\sum_{i=0}^{q-2} \alpha_i x^i + \alpha_{q-1} x^{q-1} \equiv \sum_{i=0}^{q-2} \alpha_i^P x^{(Pi)_{q-1}} + \alpha_{q-1}^P x^{q-1} \qquad (\mathrm{mod}\, x^q - x).$$

Both sides are of degree $\leqq q-1$ and therefore they are equal to one another. Taking into consideration that $i \to (Pi)_{q-1}$ is a permutation of the set $\{0, \ldots, q-2\}$, we infer by comparison of the coefficients that

$$\alpha_{(Pi)_{q-1}} = \alpha_i^P \qquad (i = 0, \ldots, q-2); \quad \alpha_{q-1} = \alpha_{q-1}^P.$$

Since, however, $(Pi)_{q-1} \equiv Pi (\mathrm{mod}\ q-1)$, it is just the condition (3) that we have obtained. Thus we have proved Theorem 2.

The conjugacy conditions (3) form a system of $q$ equations with $q$ unknowns $\alpha_0, \ldots, \alpha_{q-1}$. Our next task consists in solving this system of equations, which will give us the explicit form of the $(\mathsf{F}, \mathsf{F}_P)$-polynomials. We first introduce some easy concepts, which will also be used later.

We introduce the *relative degree*

$$m = (\mathsf{F}|\mathsf{F}_P)^{\circ}, \tag{5}$$

for which, consequently,

$$q = P^m \tag{6}$$

holds. We write the $P$-adic development of the elements $i$ of the set $0, \ldots, q-1$ in the form

$$i = i_0 + i_1 P + \cdots + i_{m-1} P^{m-1} \qquad (i_0, \ldots, i_{m-1} = 0, \ldots, P-1), \tag{7}$$

where we have considered all the $m$ numerals (also the vanishing ones). By $\sqcap$ we denote the cyclic permutation of the numerals, defined by

$$i^{\sqcap} = i_{m-1} + i_0 P + \cdots + i_{m-2} P^{m-1}. \tag{8}$$

We call $\sqcap$ the *cyclic numeral-permutation* (belonging to $P$) for the elements of the set $\{0, \ldots, q-1\}$. (It is clear that $i \to i^{\sqcap}$ supplies us with a permutation of this set which, however, is no longer cyclic; it even always possesses fixed elements, to be discussed later). The smallest natural number $l$ with

$$i^{\sqcap^l} = i$$

is called the ($P$-*adic*) *period* of $i$, which we denote here by $l(i)\,(=l_P(i))$; furthermore, we shall call the set

$$\{i, i^{\sqcap}, \ldots, i^{\sqcap^{l(i)-1}}\} \tag{9}$$

a *cycle* (more precisely a $P$-cycle) of the set $\{0, \ldots, q-1\}$. Evidently, we have $l(i)|m$. It is clear that the cycles constitute a partition of this set and the elements of a cycle will therefore be said to be *cyclically equivalent*. They are, of course, of the same $P$-adic height, whence it follows that for every $k=0, \ldots$ $\ldots, P-1$ the set of the elements of $\{0, \ldots, q-1\}$ of the $P$-adic height $k$ is a union of cycles (which, consequently, may likewise be called the cycles of the last-mentioned set). The multiples

$$k\frac{q-1}{P-1} (= k + kP + \cdots + kP^{m-1}) \qquad (k = 0, \ldots, P-1) \tag{10}$$

of $\dfrac{q-1}{P-1}$ are given by the totality of the elements of $0, \ldots, q-1$ with equal numerals (i.e. with the period length 1); for this reason these and only these elements constitute a cycle by themselves. In other terms: the elements (10) are the only fixed elements of the permutation $i \to i^{\sqcap}$ of $\{0, \ldots, q-1\}$.

We prove

Proposition 6. *The conjugacy conditions* (3) *can be written in the form*

$$\alpha_{i^{\sqcap}} = \alpha_i^P \qquad (i = 0, \ldots, q-1). \tag{11}$$

From (7) and (8) follows

$$Pi - i^{\sqcap} = (P^m - 1)\, i_{m-1}$$

and thus on account of (6) we have

$$Pi \equiv i^{\sqcap} \qquad (\operatorname{mod} q-1). \tag{12}$$

As $(q-1)^{\sqcap} = q-1$, the last equation in (3) agrees with equation (11), corresponding to $i = q-1$. It remains to be proved that the equations of the systems (3) and (11), corresponding to $i \in \{0, \ldots, q-2\}$, likewise agree with each other. Since $i^{\sqcap} \in \{0, \ldots, q-2\}$, it follows from (12) that the $j$ occurring in (3) is equal to $i^{\sqcap}$. Thus we have proved Proposition 6.

An $(\mathsf{F}, \mathsf{F}_P)$-polynomial will be called *minimal* if it is different from 0, it has a degree $\leqq q-1$ and none of its proper partial sums is an $(\mathsf{F}, \mathsf{F}_P)$-polynomial. The minimal $(\mathsf{F}, \mathsf{F}_P)$-polynomials, and hence all the $(\mathsf{F}, \mathsf{F}_P)$-polynomials, too, can be given by their explicit coefficient representation, as follows:

THEOREM 3. *The exponent sets of the minimal* $(\mathsf{F}, \mathsf{F}_P)$*-polynomials are the P-cycles of the set* $\{0, \ldots, q-1\}$. *For such a P-cycle* $\mathfrak{Z}$ *all the (minimal)* $(\mathsf{F}, \mathsf{F}_P)$*-polynomials with exponent set* $\mathfrak{Z}$ *are obtained in the form*

$$f_{\mathfrak{Z}}(x;\alpha) = \sum_{k=0}^{O(\mathfrak{Z})-1} \alpha^{P^k} x^{i^{\sqcap^k}} \qquad (\alpha \in \mathsf{F}_{P^{O(\mathfrak{Z})}} \setminus 0), \tag{13}$$

*where i is an arbitrarily chosen fixed representative of* $\mathfrak{Z}$. *One obtains all the different* $(\mathsf{F}, \mathsf{F}_P)$*-polynomials of degree* $\leqq q-1$ *by forming sums of polynomials* $f_{\mathfrak{Z}}(x;\alpha)$ *of this kind, belonging to different cycles* $\mathfrak{Z}$.

Let us call those $f_{\mathfrak{Z}}(x;\alpha)$ which occur in the (unique) decomposition described here of an $(\mathsf{F}, \mathsf{F}_P)$-polynomial of degree $\leqq q-1$ the *minimal constituents*, belonging to the single cycles $\mathfrak{Z}$, of an $(\mathsf{F}, \mathsf{F}_P)$-polynomial. It should be stressed that there is no need for each $\mathfrak{Z}$ to have a minimal constituent. We agree, for the sake of convenience, to admit in the applications also vanishing minimal constituents, in order to ensure that an $(\mathsf{F}, \mathsf{F}_P)$-polynomial of degree $\leqq q-1$ should possess exactly one minimal constituent for each cycle.

The proof of Theorem 3 is easy. Let us consider all $(\mathsf{F}, \mathsf{F}_P)$-polynomials $f(x)$ with $f^{\circ} \leqq q-1$, writing them in the form (2) where, according to Theorem 2 and Proposition 6, all solutions of the system of equations (11), to be denoted here by $S$, are admitted as coefficients $\alpha_0, \ldots, \alpha_{q-1}$. For a cycle $\mathfrak{Z}$ we denote by $S_{\mathfrak{Z}}$ the subsystem of $S$ consisting of the equations

$$\alpha_{i^{\sqcap}} = \alpha_i^P \qquad (i \in \mathfrak{Z}). \tag{14}$$

Since here $i^{\sqcap} \in \mathfrak{Z}$, those coefficients $\alpha_0, \ldots, \alpha_{q-1}$ occur in this $S_{\mathfrak{Z}}$ whose suffices lie in $\mathfrak{Z}$. (It can be said therefore that the systems $S_{\mathfrak{Z}}$ are independent of each other.)

It can easily be shown by induction that $S_{\mathfrak{Z}}$ is equivalent to the (infinite) system of equations

$$\alpha_{i^{\sqcap^k}} = \alpha_i^{P^k} \qquad (i \in \mathfrak{Z};\ k = 1, 2, \ldots). \tag{15}$$

It suffices here to admit merely an arbitrarily chosen fixed $i \in \mathfrak{z}$. For

$$k = l_P(i) (= O(\mathfrak{z}))$$

the left-hand side of (15) is equal to $\alpha_i$, whence it follows that in (15)

$$\alpha_i \in \mathsf{F}_{P^{O(\mathfrak{z})}} \tag{16}$$

must hold. Conversely, if (16) is satisfied, it is sufficient to extend the equations (15) (with a fixed $i \in \mathfrak{z}$) to the values $k = 1, \ldots, O(\mathfrak{z}) - 1$. These equations can be considered as explicit solution formulae for the system $S_{\mathfrak{z}}$, where all the unknowns are expressed by $\alpha_i$ and this $\alpha_i$ is subjected to nothing but the condition (16). After substitution (2) turns into

$$f(x) = \sum_{\mathfrak{z} \subseteqq \{0, \ldots, q-1\}} \sum_{k=0}^{O(\mathfrak{z})-1} \alpha_i^{P^k} x^{i \sqcap k}, \tag{17}$$

where—as has already been pointed out—$i$ denotes each time an arbitrary representative of the cycle $\mathfrak{z}$ and $\alpha_i$ is subjected (only) to the condition (16) (and therefore the values of the "parameter" $\alpha_i$, occurring in (17), can be chosen independently of one another). It is easily seen that the inner sum in (17) vanishes for $\alpha_i = 0$, while for $\alpha_i \neq 0$ it passes over into the right-hand side of (13); this proves Theorem 3.

NOTE 1. Though it can be stated that Theorems 1, 2 are outstripped by Theorem 3, they are often used with considerable advantage,

NOTE 2. As is well-known, for every function

$$\xi \to \xi^* \qquad (\xi, \xi^* \in \mathsf{F})$$

there exists exactly one polynomial $f(x) \in \mathsf{F}[x]$ with $f^\circ \leqq q-1$ and

$$f(\xi) = \xi^* \qquad (\xi \in \mathsf{F});$$

moreover, this polynomial is given by the interpolation formula

$$f(x) = \sum_{\xi \leqq \mathsf{F}} \xi^* (1 - (x - \xi)^{q-1}). \tag{18}$$

Thus all $(\mathsf{F}, \mathsf{F}_P)$-polynomials $f(x)$ of degree $\leqq q-1$ are represented by (18) under the restriction $\xi^* \in \mathsf{F}_P$, whence it follows that the number of these polynomials is equal to

$$P^q. \tag{19}$$

Because of its composite character, formula (18) cannot be usefully employed for the deduction of the above theorems.

NOTE 3. According to Theorem 3, all the $(\mathsf{F}, \mathsf{F}_P)$-monomials of degree $\leqq q-1$ are given by

$$\alpha x^{k\frac{q-1}{P-1}} \qquad (\alpha \in \mathsf{F}_P \setminus 0;\ k = 0, \ldots, P-1),$$

a fact which can also be seen immediately.

NOTE 4. For an $(\mathsf{F}, \mathsf{F}_P)$-polynomial $f(x) \neq 0$ a polynomial $\varrho f(x)$ $(\varrho \in \mathsf{F})$, associated with it, is an $(\mathsf{F}, \mathsf{F}_P)$-polynomial if and only if $\varrho \in \mathsf{F}_P$. It follows from this that any investigation of polynomials with restricted range of values cannot, in general, be confined to monic polynomials.

NOTE 5. Since the elements of a $P$-cycle and the $P$-cycle itself are of equal $P$-adic height, it follows from Theorem 3 that the $(\mathsf{F}, \mathsf{F}_P)$-polynomials $f(x)$ with $f^\circ \leqq q-1$ and $f^{\circ_P} \leqq k$ can simply be obtained by forming sums of minimal $(\mathsf{F}, \mathsf{F}_P)$-polynomials whose exponent sets are ($P$-cycles) of a $P$-adic height $\leqq k$.

NOTE 6. Let $\Phi, \Phi_1, \ldots, \Phi_r$ be elements of a polynomial ring free of zero divisors. (In our applications this ring will always be $\mathsf{F}[x]$.) A sum decomposition

$$\Phi = \Phi_1 + \cdots + \Phi_r$$

is called *direct* if on the right-hand side (after addition) no contraction of terms is possible, i.e. if the number of the terms of $\Phi$ is equal to the sum of the numbers of the terms of every $\Phi_i$. (In this definition the property of being free of zero divisors is not essential.) Similarly, we call a product decomposition *direct* and use for it the notation

$$\Phi = \Phi_1 \times \ldots \times \Phi_r$$

if on the right-hand side (after having carried out the multiplications) no contraction of terms is possible, i.e. if the number of the terms of $\Phi$ is equal to the product of the numbers of the terms of the factors. (This is to be understood in such a way that, in particular, each vanishing product is direct.) Direct sums of polynomials will often occur and in most cases we will not call the reader's attention to this circumstance. Concerning Theorem 3 it should be noted, however, that according to it every $(\mathsf{F}, \mathsf{F}_P)$-polynomial is the direct sum of its minimal constituents. Certain direct products of polynomials will appear later.

NOTE 7. Polynomials with a restricted range of values can be defined for infinite fundamental fields in a similar way. They rarely occur (Rédei [14]).

## § 4. Stem polynomials

For each $P \underset{1}{|} q$ we call a polynomial $f(x)$ with

$$f(x) \underset{1}{\in} \mathsf{F}[x], \quad f^\circ = \frac{q-1}{P-1}, \qquad f^{\circ_P} = 1, \quad x^q - x \,|\, f(x)^P - f(x) \tag{1}$$

a *stem polynomial*, belonging to $P$ (or a stem polynomial for $P$). We have chosen this name because certain kinds of fully reducible parallel stem polynomials will play an important role in connection with our main problem (and we could not find a better term for them).

NOTE. Since, by virtue of Theorem 1, $(1_4)$ signifies that $f(x)$ is an $(\mathsf{F}, \mathsf{F}_P)$-polynomial, the stem polynomials are the monic $P$-linear $(\mathsf{F}, \mathsf{F}_P)$-polynomials of degree $\frac{q-1}{P-1}$. Furthermore, they are completely known, on account of Theorem 3. In spite of this we shall adhere to the above definition in our discussions and we shall rarely use the explicit representation of the stem polynomials. We shall give a more exact account of this topic in § 6.

EXAMPLE. In the case $P^4 = q$, i.e. $(\mathsf{F}|\mathsf{F}_P)^\circ = 4$, the stem polynomials, belonging to $P$, are (on account of Theorems 1, 3) the

$$\begin{aligned} f(x) = {} & x^{P^3+P^2+P+1} + \\ & + \alpha x^{P^2+P+1} + \alpha^P x^{P^3+P^2+P} + \alpha^{P^2} x^{P^3+P^2+1} + \alpha^{P^3} x^{P^3+P+1} + \\ & + \beta x^{P+1} + \beta^P x^{P^2+P} + \beta^{P^2} x^{P^3+P^2} + \beta^{P^3} x^{P^3+1} + \\ & + \gamma x^{P^2+1} + \gamma^P x^{P^3+P} + \\ & + \delta x + \delta^P x^P + \delta^{P^2} x^{P^2} + \delta^{P^3} x^{P^3} + \\ & + \varepsilon \end{aligned}$$

with

$$\alpha, \beta, \delta \in \mathsf{F}(= \mathsf{F}_q = \mathsf{F}_{P^4}), \quad \gamma \in \mathsf{F}_{P^2}, \quad \varepsilon \in \mathsf{F}_P.$$

## § 5. Various examples of lacunary polynomials

As a preliminary, but also of importance later on, we define the *(generalized) binomial coefficient*, as usual, by

$$\binom{\zeta}{k} = \begin{cases} \dfrac{\zeta(\zeta-1)\dots(\zeta-k+1)}{k!} & (k = 0, 1, \dots), \\ 0 & (k = -1, -2, \dots), \end{cases}$$

where $\zeta$ may be any element of a field, containing the rational numbers. In particular, we have $\binom{\zeta}{0} = 1$; moreover, the (unrestricted) *addition theorem*

$$\binom{\zeta}{k} = \binom{\zeta-1}{k} + \binom{\zeta-1}{k-1} \tag{1}$$

and the *reduction formula*

$$\binom{\zeta}{k} = \frac{\zeta}{k}\binom{\zeta-1}{k-1} \qquad (k \neq 0) \tag{2}$$

hold.

PROPOSITION 7. *If $r$ is a rational $p$-integer, then $\binom{r}{k}$ is also a $p$-integer.*

This assertion is correct for $k \leqq 0$. In the case $k > 0$ we assume its validity for $k-1$. Since the second term on the right-hand side of (1) is then a $p$-integer, the assertion is invariant with respect to the substitution $r \to r-1$. The same thing follows by repeated application for every substitution $r \to r-l$ ($l=1, 2, \ldots$). Let $l$ be chosen in such a way that

$$\frac{r-l}{k}$$

is integer for $p$. It then follows from (2) and the induction hypothesis that the assertion is correct for $r-l$ instead of $r$, consequently it is generally valid. This proves Proposition 7.

As a result of this proposition, in the case of a rational $p$-integer $r$ the binomial coefficient $\binom{r}{k}$ can be thought of as an element of $\mathsf{F}_p$. We shall do so tacitly or with the hint in "$\mathsf{F}_p$".

PROPOSITION 8. *If $k \in \mathcal{N}$ and $r$ is a rational $p$-integer, and if the $P$-adic expansions*

$$k = k_0 + k_1 P + \cdots + k_l P^l, \quad r = r_0 + r_1 P + \cdots \qquad (k_i, r_j = 0, \ldots, P-1)$$

*are valid then the rule*

$$\binom{r}{k} = \binom{r_0}{k_0}\binom{r_1}{k_1} \cdots \binom{r_l}{k_l}$$

*holds.*

This proposition is an easy generalization of Lucas' theorem. We shall give a short proof of it, in the ring $\mathsf{F}[[x]]$. In the latter we have

$$(1+x)^r = \prod_{0 \leqq i < \infty} (1+x)^{r_i P^i} = \prod_{0 \leqq i < \infty} (1+x^{P^i})^{r_i},$$

i.e.

$$\sum_{0 \leqq k < \infty} \binom{r}{k} x^k \left( = \prod_{0 \leqq i < \infty} \sum_{j_i=0}^{P-1} \binom{r_i}{j_i} x^{j_i P^i} \right) = \sum_{j_0, j_1, \ldots = 0}^{P-1} \binom{r_0}{j_0} \binom{r_1}{j_1} \ldots x^{j_0 + j_1 P + \ldots}.$$

Hence Proposition 8 follows by the comparison of the coefficients.

We shall now discuss some lacunary polynomials in $\mathsf{F}[x]$, arising in various connections. These examples will show us that in $\mathsf{F}[x]$ we very often have to deal with such polynomials, also without special intention.

It can be inferred from the case $P=p$ of Proposition 8 that the powers

$$(x+\alpha)^k \qquad (\alpha \in \mathsf{F},\ k \in \mathcal{N};\ k \neq 1, 2, \ldots, p-1, p^2-1, p^3-1, \ldots)$$

are lacunary.

The derivative of an $f(x) \in \mathsf{F}[x]$ is lacunary in the case $f^\circ \geqq p+1$, since the exponent set $\mathfrak{E}(f')$ contains only elements $\not\equiv -1 \pmod p$.

We have pointed out already that the polynomials $f(x) \in \mathsf{F}[x]$ with $f'(x)=0$ are lacunary, since for these the exponent set $\mathfrak{E}(f)$ has only elements divisible by $p$.

As rather important examples of lacunary polynomials, we mention the Galois binomial and the Euler binomials, both fully reducible, as already noted above.

## § 6. Problems I, II, III concerning fully reducible lacunary polynomials

According to what has been stated in the preface, our main task consists in the determination of certain fully reducible lacunary polynomials in $\mathsf{F}[x]$. Of course, we can only expect to find partial solutions to this problem. It is best to investigate not only the reducibility of single lacunary polynomials but to categorise certain lacunary polynomials as a 'type", endeavouring then to determine which polynomials of a given type are fully reducible ones. It is advisable to admit, for each type, only polynomials of equal degree and therefore we agree on the following definition.

For each finite set $\mathfrak{L}$ of non-negative integer numbers, neither empty nor of the form $\{0, 1, \ldots, k\}$, we denote by $\mathfrak{L}_x$ the (finite) set of the polynomials $f(x) \in \mathsf{F}[x]$, satisfying the conditions

$$f^\circ = \max \mathfrak{L}, \quad \mathfrak{E}(f) \subseteqq \mathfrak{L};$$

we call $\mathfrak{L}_x$ the *lacunarity type* belonging to $\mathfrak{L}$. Then by the *$\mathfrak{L}$-problem* we understand the determination of the fully reducible polynomials in $\mathfrak{L}_x$.

We shall be concerned with certain special cases of this still very general problem which we will call Problems I, II, III (as indicated in the preface).

PROBLEM I. *For every $d|q-1$ ($d>1$) to determine those polynomials*

$$f(x) \underset{1}{\in} \mathsf{F}[x] \setminus x\mathsf{F}[x] \tag{1}$$

*with*

$$f^{\circ} = \frac{q-1}{d}, \quad f^{\circ\circ} \leqq \frac{q-1}{d^2} \tag{2}$$

*which are fully reducible and possess no multiple factors.*

(Of course, (1) is to be understood as $f(x) \underset{1}{\in} \mathsf{F}[x]$ and $f(x) \notin x\mathsf{F}[x]$.)

PROBLEM II. *Among the polynomials*

$$f(x) \underset{1}{\in} \mathsf{F}[x] \setminus \mathsf{F}[x^p] \tag{3}$$

*with*

$$f^{\circ} = q, \quad f^{\circ\circ} \leqq \frac{q+1}{2} \tag{4}$$

*to determine those that are fully reducible.*

The reader might observe that these two problems are (outwardly) rather similar, the similarity being reflected in the conditions (2) and (4) (each establishing a lacunarity type). However, we had to exclude from Problem I the $f(x)$ divisible by $x$ and those with multiple factors, and from Problem II the degenerate $f(x)$, i.e. those with $f'(x)=0$. Without these restrictions there seems to be no prospect of the determination of all solutions. (In § 33 we shall investigate the question of the degenerate solutions of Problem II.) "Inwardly" both problems are very different because Problem II is quite unexpectedly difficult and complex (see below), whereas Problem I will provide no serious difficulties. By way of compensation, Problem II is much more important for the applications than Problem I, and therefore its solution is to be considered our most important task. We shall sometimes call it our "main problem".

We are unable to solve Problem II without reformulating it several times; in this way we are led to some further problems, interesting in their own right, too, which will therefore occasionally be called the *auxiliary problems* "induced" by Problem II. In this series of (successive) auxiliary problems the last is perhaps the most interesting, concerning the stem polynomials (belonging to a given $P$), i.e. (cf. § 4) the $f(x)$ with the property

$$f(x) \underset{1}{\in} \mathsf{F}[x], \quad f^\circ = \frac{q-1}{P-1}, \quad f^{\circ_P} = 1, \quad x^q - x | f(x)^P - f(x). \tag{5}$$

being again "essentially" a (rather complex) $\mathfrak{L}$-problem, namely:

PROBLEM III. *In the case $p \neq 2$ for each pair $P$, $\varphi(x)$ with*

$$P \underset{1}{|} q \tag{6}$$

*and*

$$\varphi(x) \underset{1}{\in} \mathsf{F}[x], \quad \varphi(x) | x^P - x, \quad \frac{P-1}{2} \leqq \varphi^\circ < P \tag{7}$$

*to determine those of the stem polynomials $f(x)$ (see (5)), belonging to $P$, that satisfy the condition*

$$\varphi(f(x)) | x^q - x. \tag{8}$$

Let such an $f(x)$ be called a *solution* of Problem III, *belonging* to the pair $P$, $\varphi(x)$. Already at this point let it be noted once and for all that in view of $(5_2)$ $P$ is unambiguously determined by $f(x)$ or even already by its degree $f^\circ$.

In order to put this in its proper light, let us notice, first of all, that $x^P - x$ is a fully reducible binomial over $\mathsf{F}_P$ and that it possesses no multiple zeros. Accordingly it follows from (7) that all the possible $\varphi(x)$ agree with the products

$$\varphi(x) = (x - \varrho_1) \dots (x - \varrho_k)$$

$$\left(\frac{P-1}{2} \leqq k < P;\ \varrho_1, \dots, \varrho_k \text{ different elements from } \mathsf{F}_P\right).$$

Therefore the condition of divisibility (8) becomes

$$(f(x) - \varrho_1) \dots (f(x) - \varrho_k) | x^q - x.$$

Now, on the one hand, the factors of the left-hand side are (different) polynomials, parallel to $f(x)$ (consequently also to one another), which are, because $\varrho_1, \dots, \varrho_k \in \mathsf{F}_P$, also stem polynomials (belonging to $P$). On the other hand, different parallel polynomials are always prime to one another, whence the above divisibility is equivalent to the totality of the divisibilities

$$f(x) - \varrho_1, \dots, f(x) - \varrho_k | x^q - x.$$

Thus Problem III can be treated as a *simultaneous $\mathfrak{L}$-problem*, requiring that simultaneously all members of a series

$$f(x) - \varrho_1, \dots, f(x) - \varrho_k \quad \left(\frac{P-1}{2} \leqq k < P\right)$$

of (different) parallel stem polynomials should be fully reducible and without multiple factors ($\varphi(x)$ being no longer mentioned here).

Although Problem III was set in its proper light only by this equivalent redrafting, we shall always treat it in its original form which is more suitable for the determination of the solutions.

It is a formal similarity of Problems I, II and III that a single sign $\leqq$ or $\geqq$ occurs in the premises of each of them (viz. in (2), (4), (7)). When $=$ is actually required, we shall refer to the *marginal case* of these problems. (Of course, for Problems II the marginal case occurs only for $p \neq 2$.)

We shall find that for Problems I, II the marginal case is of particular importance. For Problem III the same thing holds even to an increased degree, as is seen from the following simple observation. Let us consider Problem III for a further pair $P$, $\psi(x)$ with $\psi(x)|\varphi(x)$ (where $\psi^\circ \geqq \frac{P-1}{2}$ must hold, of course). Every solution of Problem III, belonging to $P$, $\varphi(x)$, also then trivially belongs to $P$, $\psi(x)$, whence it follows that all its solutions hold for the marginal case too. It was therefore essentially sufficient to consider Problem III only for the marginal case; we abide, however, by the above degree of generality, because in this way we shall be able to give a better survey of the solutions, without any additional trouble. We must add to this remark the following definition: A solution of Problem III is said to be *strong* if it belongs to (at least) one pair $P$, $\varphi(x)$ for which $\varphi^\circ \geqq \frac{P+1}{2}$ holds. The remaining solutions are called *weak*. [Thus the weak solutions belong to a single pair $P$, $\varphi(x)$ and in this pair $\varphi^\circ = \frac{P-1}{2}$ holds, while the strong ones belong to at least two different pairs $P$, $\varphi(x)$ and $P$, $\psi(x)$.]

We shall solve Problem I completely and Problem II "almost" completely. Exceptional cases in which, apart from the solutions to be given by us, some further ones may possibly exist, can occur—as we shall see—only in the marginal case of Problem II and even then only for $q=3^n$, $5^n$, $7^n$ $(n \geqq 4)$. We conjecture that in reality there are no exceptional cases.

In the solution of Problem III we have not striven for completeness, this problem being, as stated above, merely an auxiliary problem to Problem II. To be perfectly honest, we should dearly have liked to give a complete solution of this problem, in view of its indisputable independent interest, but we have not succeeded in doing so, on account of its difficulty. Nevertheless, we have been able to carry the investigations far enough to make the success of further researches probable. We shall determine the strong solutions entirely. As to the weak solutions, we shall content ourselves with partial results, making possible the successful application to Problem II.

It should be emphasized that both Problems I and II have only a few solutions not belonging to the marginal case; this fact will be advantageous for applications.

We still wish to elucidate more exactly the relationship between Problems II and III. This relationship really only concerns the marginal cases of both problems, stating, as we shall see, that in order to obtain all solutions (for the marginal case) of Problem II, it would be sufficient to know the solutions of Problem III for every $p \neq 2$, belonging to the unique pair

$$P = p, \quad \varphi(x) = x^{\frac{P-1}{2}} + 1.$$

We were nevertheless driven to treat Problem III in much greater generality, because otherwise the very complicated procedure of induction necessary for mastering it would not be possible.

NOTE 1. For any arbitrary lacunarity type

$$\mathfrak{L} = \{k_0, \ldots, k_t\} \qquad (k = k_0 > k_1 > \cdots > k_t \geqq 0)$$

the $\mathfrak{L}$-problem admits of the following trivial redrafting. All solutions are represented by the products

$$\prod_{i=1}^{k} (x - \varrho_i)$$

in which for $\varrho_1, \ldots, \varrho_k$ all solutions of the system of equations

$$s_l(\varrho_1, \ldots, \varrho_k) = 0 \qquad (1 \leqq l \leqq k;\ l \neq k - k_1, \ldots, k - k_t), \tag{9}$$

lying in $\mathsf{F}$, are to be admitted, where the left-hand side denotes the $l$-th elementary symmetric expression of $\varrho_1, \ldots, \varrho_k$. Although we shall need this equivalent redrafting for the purpose of the applications, it yields no utilizable procedure for the solution of the $\mathfrak{L}$-problem, since the system (9) is in general far too complicated.

NOTE 2. We shall make an assault on all three Problems I, II, III by means of a device which appears to be indispensable.

NOTE 3. In all three problems we are concerned with monic polynomials $f(x)$. This restriction is evidently not essential for Problems I and II and the same thing holds for Problem III, as is easily seen from Theorem 3.

## § 7. The $\lambda$-differential equation and the $a$, $b$-polynomial equation

By the term *polynomial equation* in $\mathsf{F}[x]$ we shall understand an equation in which nothing but polynomials from $\mathsf{F}[x]$ occur, some of which are allowed to be unknown. If derivatives of the unknown polynomials also occur in the equation, we shall refer to a *differential equation* in $\mathsf{F}[x]$.

Of course one very often has to deal with polynomial equations in $\mathsf{F}[x]$. For instance, a condition of divisibility $f(x)|g(x)$ can be turned into the polynomial equation $f(x)u(x)=g(x)$, containing the unknown polynomial $u(x)$. In a similar manner we could also change Problems I, II, III into polynomial equations (with certain accessory conditions), but that would be pointless.

Although all kinds of polynomial equations are in principle solvable by comparison of the coefficients, such a solution in most cases involves enormously difficult calculation. This makes it advisable to look for other procedures. We shall return to this point later.

As for the differential equations in $\mathsf{F}[x]$, we note that their treatment is in general particularly difficult; this can be explained by the fact that the solutions of the differential equation $f'(x)=0$ appear in the form

$$f(x) = \alpha_0+\alpha_1 x^p+\cdots+\alpha_k x^{pk} \qquad (\alpha_0, \ldots, \alpha_k \in \mathsf{F}),$$

from which we see that arbitrarily many "integration constants" occur here. We therefore have to reckon with the possibility of similar difficulties in other cases of differential equations in $\mathsf{F}[x]$. As in most cases we have to deal with differential equations with "accessory conditions", it is clear that the determination of the appropriate values of the integration constants may be very difficult.

In the course of our discussion we shall meet several "ordinary" polynomial equations as well as differential equations in $\mathsf{F}[x]$. Here we shall mention only two such equations, which occur in connection with (the marginal case of) Problem II, the solution of which will involve a considerable amount of work. These are, for $p\neq 2$, the equations

$$(g(x)^2)' = x^q-x+\lambda g(x) \qquad \left(\lambda\in\mathsf{F};\ g(x)\underset{1}{\in}\mathsf{F}[x];\ g^\circ = \frac{q-1}{2}\right) \tag{1}$$

and

$$xa(x)^p+b(x)^p = x^r a(x)^{p-2}+a(x)^{p-3}b(x) \tag{2}$$

$$\left(r = \frac{q}{p};\ a(x)\underset{1}{\in}\mathsf{F}[x],\ b(x)\in\mathsf{F}[x];\ b^\circ \leqq a^\circ \leqq \frac{r-p}{2}\right).$$

We shall call them the *λ-differential equation* and the *a, b-polynomial equation*, respectively, for short. (The first is of the first order with the arbitrary (constant) parameter $\lambda$. The second contains the two unknowns $a(x)$, $b(x)$.) Equation (1) will be needed directly for Problem II, while (2) will be needed for (1).

We wish to make the following remarks in connection with these two equations in advance.

Equation (1) is solvable only for the two values $\lambda = \pm 3$ of the parameter (i.e. when $p=3$ only for $\lambda=0$). We shall arrange the proof so as to yield at the same time also the solutions of (1), however not only the required ones, but also those lying in $\mathsf{F}(x) \setminus \mathsf{F}[x]$. It will give us a great deal of trouble to find those that do lie in $\mathsf{F}[x]$.

In order to be able to deal with equation (2), we have to admit certain power series as solutions. In this way the "uniformization" of (2) becomes possible, but then we must select the suitable solutions (i.e. those lying in $\mathsf{F}[x]$). This too will turn out to be very difficult.

Finally, we remark that the solutions $g(x)$ of (1) will be lacunary polynomials, containing only exponents which are congruent mod $p$ to one of the numbers $0, \ldots, \frac{p+1}{2}$. The solutions $a(x)$, $b(x)$ of (2) will also be closely connected with lacunary polynomials, inasmuch as it will be possible to represent them by certain $P$-linear polynomials. This was quite unforeseen, because it is not immediately obvious from equations (1) and (2) that they will lead to lacunary polynomials.

It should be emphasized that, according to what we have said above, both equations (1) and (2) will be treated by application of the principle of the extension of the domain of solutions.

## § 8. Fully reducible binomials

The Euler binomials will play an important role in connection with Problems I, II and III. We already have pointed out that they are fully reducible. For the sake of completeness, we propose to determine all fully reducible binomials in $\mathsf{F}[x]$, which will provide us with the simplest cases of fully reducible lacunary polynomials. It is, of course, sufficient to take into account merely the monic binomials.

THEOREM 4. *All reducible monic binomials from* $\mathsf{F}[x]$ *are given uniquely by*

$$x^i f(x)^{pj} \qquad (i, j = 0, 1, \ldots; f(x) \text{ an Euler binomial}). \tag{1}$$

It is clear that the polynomials (1) are monic binomials and that they are fully reducible. Conversely, suppose that we are given a fully reducible binomial

$$g(x) \underset{1}{\in} \mathsf{F}[x].$$

This binomial can be written (initially regardless of its fully reducibility) in the form

$$g(x) = x^i(x^k-\alpha)^{p^j} \qquad (i, j = 0, 1, \dots; p \nmid k;\ \alpha \in \mathsf{F} \setminus 0),$$

where $i$, $j$, $k$, $\alpha$ are uniquely determined. As a consequence of the full reducibility of $g(x)$, it follows that the binomial $x^k-\alpha$ must possess the same property. On the other hand (taking into consideration that $p \nmid k$), let us calculate the greatest common divisor

$$\big(x^k-\alpha,\ (x^k-\alpha)'\big) = (x^k-\alpha,\ kx^{k-1}) = 1.$$

Accordingly, $x^k-\alpha$ has no multiple zeros. This, when combined with the previous result, implies that

$$x^k-\alpha | x^q-x.$$

Because

$$\alpha \neq 0$$

even

$$x^k-\alpha | x^{q-1}-1$$

holds, and thus

$$k|q-1, \quad \alpha^{\frac{q-1}{k}} = 1.$$

Since, however, the group $\mathsf{F} \setminus 0$ is cyclic and of the order $q-1$, we may further infer that

$$\alpha \in \mathsf{F}^k \setminus 0.$$

Consequently $x^k-\alpha$ is an Euler binomial. Thus we have proved Theorem 4.

We now proceed to the proof of the following

SUPPLEMENT. *For every divisor $k|q-1$ all Euler binomials of degree $k$ are equivalent to the Euler binomial $x^k-1$.*

An Euler binomial of degree $k$ can be written in the form

$$x^k-\varrho^k \qquad (\varrho \in \mathsf{F} \setminus 0).$$

Hence and from

$$\varrho^{-k}\big((\varrho x)^k-\varrho^k\big) = x^k-1$$

the supplement follows.

Let $x^k-1$ $(k|q-1)$ be called the *special Euler binomial* of degree $k$. This binomial lies in $\mathsf{F}_p[x]$.

NOTE 1. If in the $\mathfrak{L}$-problem we restrict ourselves to the case $\mathfrak{L} \subset \{0, \ldots, q-2\}$, then the question concerns polynomials of the form

$$f(x) = \alpha_0 + \alpha_1 x + \cdots + \alpha_{q-2} x^{q-2} \qquad (\alpha_0, \ldots, \alpha_{q-2} \in \mathsf{F}).$$

If $r$ denotes the rank of the cyclic matrix

$$\begin{pmatrix} \alpha_0 & \alpha_1 \cdots \alpha_{q-2} \\ \alpha_{q-2} & \alpha_0 \cdots \alpha_{q-3} \\ \vdots & \\ \alpha_1 & \alpha_2 \cdots \alpha_0 \end{pmatrix},$$

then it follows from the generalized theorem of König–Rados (see Rédei [9]) that $f(x)$ has exactly $q-1-r$ zeros in $\mathsf{F}$ which are different from one another and from 0. Concerning a similar theorem see Rédei–Turán [10]. In simple cases both theorems seem to be applicable to the determination of the fully reducible $f(x)$, but success cannot easily be attained in this way. In the present book we shall not use these theorems.

NOTE 2. According to a part of a theorem of our own (Rédei [5] or [9]) the cubic (lacunary) trinomial

$$x^3 + \alpha x + \beta \qquad (\alpha, \beta \in \mathsf{F} \setminus 0)$$

in the case $p \neq 2, 3$ is fully reducible and without multiple zeros if and only if the discriminant $D = -4\alpha^2 - 27\beta^3$ lies in $\mathsf{F}^2 \setminus 0$ and

$$\varphi_k\left(-\frac{27\beta^3}{D}\right) = 0 \qquad \left(k = \left[\frac{q+1}{3}\right]\right)$$

holds, where

$$\varphi_k(x) = \binom{k}{1} + \binom{k}{3} x + \binom{k}{5} x^2 + \cdots + \binom{k}{k} x^{\frac{k+1}{2}} \qquad (\in \mathsf{F}_r[x]).$$

PROBLEM (P. Turán). Determine the fully reducible trinomials in $\mathsf{F}[x]$, the degrees of which is $\geqq 4$.

CHAPTER II

# PROBLEM I

We shall succeed in solving Problem I essentially with the aid of a single device.

## § 9. The solution of Problem I

We assume that $f(x)$ is a solution of Problem I. This means, as is seen by a simple redrafting, that

$$f(x) \underset{1}{\in} \mathsf{F}[x], \quad f^{\circ} = \frac{q-1}{d}, \quad f^{\circ\circ} \leqq \frac{q-1}{d^2} \qquad (d|q-1, \; d > 1) \tag{1}$$

and

$$f(x)|x^{q-1}-1 \tag{2}$$

hold, where $d$ denotes a fixed number. Because of $(1_2)$ we have $f^{\circ} \geqq 1$ and thus, by virtue of (2), $f(x)$ must have at least two terms. The $d$ Euler binomials of degree $\frac{q-1}{d}$ are obviously solutions of Problem I, which we shall sometimes call the *trivial solutions* of Problem I. Henceforth let $f(x)$ denote a further solution. Then it follows from $(1_2)$, (2) and Theorem 4 that $f(x)$ must have three terms at least.

In order to investigate $f(x)$ in this case, let us avail ourselves of the trivial solutions $\left(\text{i.e. the Euler binomials of degree } \frac{q-1}{d}\right)$

$$f_{\alpha}(x) = x^{\frac{q-1}{d}} - \alpha \qquad \left(\alpha \in \mathsf{F}^{\frac{q-1}{d}} \setminus 0\right), \tag{3}$$

first forming the differences

$$g_{\alpha}(x) = f(x) - f_{\alpha}(x) \qquad (\in \mathsf{F}[x] \setminus \mathsf{F}) \tag{4}$$

and then their product

$$F(x) = \prod_{\alpha} g_{\alpha}(x) \left(= \prod_{\alpha} \left(f(x) - f_{\alpha}(x)\right)\right). \tag{5}$$

Here and in what follows $\alpha$ has the same meaning as in (3).

By virtue of

$$\prod_\alpha f_\alpha(x) = x^{q-1}-1$$

and (2) the divisibility relation

$$f(x)|F(x) \tag{6}$$

follows from (5), after having carried out the multiplications.

We estimate the degree $F^\circ$ of $F(x)$ from below and from above. Since $f(x)$ and $f_\alpha(x)$ are monic polynomials of equal degree, ($1_3$), (3) and (4) imply

$$0 < g_\alpha^\circ = f^{\circ\circ} \leqq \frac{q-1}{d^2}.$$

Since the number of the $\alpha$ is $d$, hence

$$0 < F^\circ = df^{\circ\circ} \leqq f^\circ$$

follows by ($1_2$), ($1_3$) and (5).

This means that both sides of (6) are (of equal degree and) associated with one another, i.e. an equation

$$F(x) = \varkappa f(x) \qquad (\varkappa \in \mathsf{F} \setminus 0) \tag{7}$$

holds.

In order to proceed, we must take into consideration that, as a result of (5), only $f(x)$ occurs as an unknown polynomial in the polynomial equation (7). It is advisable, however, to introduce the new unknown

$$h(x) = f(x) - x^{\frac{q-1}{d}} \tag{8}$$

in place of $f(x)$. For its degree, the upper estimate

$$h^\circ (= f^{\circ\circ}) \leqq \frac{q-1}{d^2} \left(= \frac{f^\circ}{d}\right) \tag{9}$$

holds, by (1).

From (3) and (8) we infer

$$f(x) - f_\alpha(x) = h(x) + \alpha.$$

Substituting this into (5) we obtain

$$F(x) = \prod_\alpha (h(x)+\alpha).$$

However, since (cf. (3))

$$\prod_\alpha (x-\alpha) = x^d - 1$$

and therefore

$$\prod_{\alpha} (x+\alpha) = x^d-(-1)^d,$$

it follows that

$$F(x) = h(x)^d-(-1)^d.$$

Because of this and of (8), (7) turns into the polynomial equation

$$h(x)^d = \varkappa\left(x^{\frac{q-1}{d}}+h(x)\right)+(-1)^d \tag{10}$$

(with the unknown $h(x)$ and the parameter $\varkappa$).

Instead of solving this equation merely by comparing coefficients, we compare the sinkings of both sides with one another. This will lead us to the result much more quickly. Since, as a result of $d|q-1$, $d$ is prime to $p$, Proposition 4 implies that

$$(h(x)^d)^{\downarrow}=h^{\downarrow}.$$

On the other hand, the sinking of the right-hand side of (10) is by (9) equal to

$$\frac{q-1}{d}-h^{\circ},$$

where we have taken into consideration that $h(x)$ has, from (8), at least two terms. We thus obtain

$$h^{\downarrow} = \frac{q-1}{d}-h^{\circ}. \tag{11}$$

Since now $h^{\downarrow}\leqq h^{\circ}$,

$$\frac{q-1}{d} = h^{\downarrow}+h^{\circ} \leqq 2h^{\circ} \leqq 2\frac{q-1}{d^2} \tag{12}$$

follows from (11) and (9). Hence $d\leqq 2$. Since, on the other hand, $d>1$, we find that

$$d=2. \tag{13}$$

Moreover, it can be also inferred that only = is possible in (12), instead of $\leqq$, and consequently that

$$h^{\circ} = \frac{q-1}{4} \tag{14}$$

and therefore that

$$4|q-1. \tag{15}$$

(In particular, we have so far found that Problem I can have solutions with at least three terms only if (13) and (15), and thus also $p\neq 2$, hold.)

In addition to this, (11), (13) and (14) imply $h^{\downarrow}=h^{\circ}$, i.e.

$$h^{\circ\circ}=0. \tag{16}$$

We shall now very soon achieve our aim. From (14) and (16), we can write

$$h(x) = \lambda x^{\frac{q-1}{4}} + \mu \qquad (\lambda, \mu \in \mathsf{F}\setminus 0). \tag{17}$$

Substituting this into (10) and taking (13) into account we obtain

$$\left(\lambda x^{\frac{q-1}{4}} + \mu\right)^2 = \varkappa \left(x^{\frac{q-1}{2}} + \lambda x^{\frac{q-1}{4}} + \mu\right) + 1.$$

Simple comparison of coefficients gives the system of equations

$$\lambda^2 = \varkappa, \quad 2\lambda\mu = \varkappa\lambda, \quad \mu^2 = \varkappa\mu + 1.$$

As $\lambda \neq 0$, the second equation implies

$$\varkappa = 2\lambda,$$

and thus the first and the third equations give

$$\lambda^2 = 2\mu, \quad \mu^2 = -1.$$

Since, accordingly, $2\mu = (1+\mu)^2$, we also have:

$$\lambda = \nu(1+\mu) \qquad (\nu^2 = 1)$$

for suitable $\nu$. By substituting this into (17)

$$h(x) = \nu(1+\mu)x^{\frac{q-1}{4}} + \mu$$

follows, and consequently

$$f(x) = x^{\frac{q-1}{2}} + \nu(1+\mu)x^{\frac{q-1}{4}} + \mu,$$

i.e.

$$f(x) = \left(x^{\frac{q-1}{4}} + \nu\right)\left(x^{\frac{q-1}{4}} + \mu\nu\right)$$

follows from (8) and (13). Here $\mu$ can be replaced by $\mu\nu$, whence (by transposition of the factors)

$$f(x) = \left(x^{\frac{q-1}{4}} + \mu\right)\left(x^{\frac{q-1}{4}} + \nu\right) \qquad (\mu^2 = -1, \ \nu^2 = 1). \tag{18}$$

Conversely, it is clear that in the case $4|q-1$ every polynomial $f(x)$ of the form (18) is a trinomial and, as the product of two different Euler binomials of equal degree, it satisfies the condition of divisibility (2). Since

$$f^{\circ} = \frac{q-1}{2} \quad \text{and} \quad f^{\circ\circ} = \frac{q-1}{4}$$

also follow from (18), the conditions (1) are likewise satisfied.

With a slight change of the notations we summarize our results as follows:

THEOREM 5. *All solutions of Problem I, i.e. the polynomials $f(x)$ with*

$$f(x) \in_1 \mathsf{F}[x], \quad f^{\circ} = \frac{q-1}{d}, \quad f^{\circ\circ} \leqq \frac{q-1}{d^2} \; (d|q-1, \; d>1), \quad f(x)|x^{q-1}-1$$

*are the d Euler binomials*

$$x^{\frac{q-1}{d}} - \alpha \qquad \left(\alpha \in \mathsf{F}^{\frac{q-1}{d}} \setminus 0\right),$$

*as well as, in the case $p \neq 2$, $4|q-1$, $d=2$, in addition the four trinomials*

$$\left(x^{\frac{q-1}{4}} - \beta\right)\left(x^{\frac{q-1}{4}} - \gamma\right) \qquad (\beta^2 = 1, \; \gamma^2 = -1).$$

The case $q=p$, $d=2$ is found in Rédei [6].

CHAPTER III

# REDUCTION OF PROBLEM II TO PROBLEM III

The "main part" of Problem II, i.e. apart from its marginal case will be solved with the aid of a simple device and the marginal case will be simultaneously reduced to the $\lambda$-differential equation. The solutions of the latter, however, will be obtained only after a long sequence of further reductions, together with the solutions of the *a*, *b*-polynomial equation. As a result of this difficult reduction process the marginal case of Problem II will be reduced to a part of the marginal case of Problem III (and it will not be solved, apart from a few exceptions, until Chapter V).

## § 10. The solution of Problem II with the exception of the marginal case and the reduction of the latter to the $\lambda$-differential equation

We mention in advance that by virtue of Proposition 1 Problem II is invariant, by which is meant that for any one of its solutions all (monic) transformed polynomials

$$\alpha^{-1}f(\alpha x+\beta) \qquad (\alpha, \beta \in \mathsf{F};\ \alpha \neq 0)$$

are also solutions of it and that, moreover, all these solutions belong to the marginal case if the same holds for $f(x)$.

Now we suppose that $f(x)$ is a solution of Problem II. This means that

$$f(x) \underset{1}{\in} \mathsf{F}[x], \quad f^{\circ} = q, \quad f^{\circ\circ} \leqq \frac{q+1}{2}, \tag{1}$$

$$f'(x) \neq 0 \tag{2}$$

and that, further, $f(x)$ is fully reducible.

It is obvious that the Galois binomial $x^q - x$ is a solution which we shall occasionally call the *trivial solution* of Problem II. (This is an "ordinary" solution, i.e. it does not belong to the marginal case.) Henceforth we assume that $f(x) \neq x^q - x$.

We again start by forming the difference

$$g(x) = f(x) - (x^q - x) \qquad (\neq 0) \tag{3}$$

of the solution $f(x)$ and the trivial solution. (1) and (3) imply that

$$g^{\circ} \leqq \frac{q+1}{2}. \tag{4}$$

In addition we avail ourselves of the zero polynomial

$$F(x) = \prod_{f(\varrho)=0} (x-\varrho)$$

of $f(x)$, where the product, as indicated, is to be extended over the different zeros of $f(x)$ (lying in $\mathsf{F}$ because $f(x)$ is fully reducible). This polynomial will play a short but important role in our considerations. Then the validity of both divisibility relations

$$F(x)|f(x),\ x^q - x,$$

as well as yet the further one

$$\left.\frac{f(x)}{F(x)}\right| f'(x)$$

can be inferred trivially (but they hold for very different reasons). Consequently both divisibilities

$$F(x)|g(x),\quad \left.\frac{f(x)}{F(x)}\right| g'(x)-1$$

follow from (3). Finally, by multiplication of the latter, the divisibility

$$f(x)|g(x)\big(g'(x)-1\big) \tag{5}$$

follows.

The left-hand side is of degree $q$, while the degree of the right-hand side is not greater than $q$, as a consequence of (4). As it also follows from (2) and (3) that the right-hand side does not vanish, it can be concluded that both sides are associated with one another and that in (4) $=$ is valid, instead of $\leqq$. Thus we first have that

$$g^{\circ} = \frac{q+1}{2}. \tag{6}$$

Accordingly $p \neq 2$ must hold. Moreover, since (5) implies $g^{\circ} \geqq 2$,

$$f^{\circ\circ} = \frac{q+1}{2}$$

arises from (3) and (6), whence it is seen that the solution $f(x)$ must belong to the marginal case of Problem II.

In order to express this property of "being associated" in a more convenient form, let us write $g(x)$ in the (unique) form

$$g(x) = \varkappa h(x) \qquad (h(x) \underset{1}{\in} \mathsf{F}[x];\ \varkappa \in \mathsf{F} \setminus 0). \tag{7}$$

It follows from this and from (6) that the coefficient of the leading term on the right-hand side of (5) is equal to $\frac{1}{2}\varkappa^2$, while that on the left-hand side is 1; consequently we have

$$\frac{1}{2}\varkappa^2 f(x) = g(x)\big(g'(x) - 1\big).$$

Substituting from (7) and dividing by $\frac{\varkappa^2}{2}$ we obtain

$$f(x)\left(= 2h(x)\left(h'(x) - \frac{1}{\varkappa}\right)\right) = \left(h(x)^2\right)' - \frac{2}{\varkappa}h(x),$$

whence, from (3) and (7),

$$\left(h(x)^2\right)' = x^q - x + \left(\varkappa + \frac{2}{\varkappa}\right)h(x).$$

Moreover, by (3), (6) and (7) we have

$$f(x) = x^q - x + \varkappa h(x), \quad h^\circ = \frac{q-1}{2}.$$

Writing $g$ again instead of $h$, we summarize the results as follows:

Proposition 9. *Apart from the trivial solution $x^q - x$ Problem II can possess further solutions only for $p \neq 2$ and these must belong to the marginal case; they are necessarily of the form*

$$f(x) = x^q - x + \varkappa g(x), \tag{8}$$

*where*

$$\left(g(x)^2\right)' = x^q - x + \left(\varkappa + \frac{2}{\varkappa}\right)g(x) \qquad \left(\varkappa \in \mathsf{F} \setminus 0;\ g(x) \in \mathsf{F}[x];\ g^\circ = \frac{q+1}{2}\right) \tag{9}$$

*holds.*

We see that (9) agrees with the case

$$\lambda = \varkappa + \frac{2}{\varkappa}$$

of the $\lambda$-differential equation; our next task therefore consists in solving the $\lambda$-differential equation. We shall admit all values of the parameter $\lambda \in \mathsf{F}$, since this greater generality does not cause any additional work.

Proposition 9 essentially gives the reduction of Problem II to the $\lambda$-differential equation. The question here is, of course, about an "incomplete" reduction, because for the determination of the solutions for the marginal case of Problem II it will be necessary, after having substituted the solutions $g(x)$ of (9) into the right-hand side of (8), to carry out the (very difficult) selection of the fully reducible polynomials $f(x)$ from among all those that this process yields. Since, on the other hand, as was pointed out earlier, the determination of the solutions of the $\lambda$-differential equation will also prove to be a very complex task, we have before us a very long program.

NOTE 1. Since (9) clearly implies $g(x)|x^q-x$, every solution $g(x)$ of (9) is necessarily fully reducible (and also without multiple divisors). On the other hand, the further divisibility $g(x)|f(x)$ follows from (8) and therefore $f(x)$ is already fully reducible if the same holds for the cofactor $g(x)^{-1}f(x)$. This advantageous state of affairs will not be exploited until much later on.

NOTE 2. It will be found that, in general, the marginal case of Problem II has many solutions and that the number of solutions increases rapidly with the characteristic $p$ as well as with the number of prime factors of the degree $n$ of the field $\mathsf{F}$.

NOTE 3. If we add 1 to the right-hand side of ($1_3$), we obtain a generalization of Problem II which can be investigated according to the above pattern; this problem leads to a somewhat more complicated differential equation and there seems no hope for a complete solution of this.

## § 11. Preliminary remarks on polynomial and differential equations

When discussing polynomial equations in $\mathsf{F}[x]$, in general we include the differential equations. Here we make a few preliminary observations which will often be useful for the solution of polynomial equations.

For an algebra $\mathsf{A}$ over a fundamental ring $\mathsf{R}$ the symbol $\mathsf{A}|\mathsf{R}$ will be used. (This is a generalization of a notation which already has been applied for field extensions.)

The principle of comparing coefficients in $\mathsf{F}[x]$ is based on the fact that $\mathsf{F}[x]|\mathsf{F}$ is an algebra with the (infinite) basis $1, x, x^2, \ldots$. Similarly, $\mathsf{F}[x]|\mathsf{F}[x^p]$ is an algebra with the (finite) basis

$$1, x, \ldots, x^{p-1}.$$

The unique basis representation corresponding to this basis:

$$f(x) = \sum_{i=0}^{p-1} x^i f_i(x^p) \qquad (f(x), f_0(x), \ldots, f_{p-1}(x) \in \mathsf{F}[x]) \tag{1}$$

is called the *component representation* of the polynomials $f(x) \in \mathsf{F}[x]$; moreover, the cofactor $f_i(x^p)$ of $x^i$ is said to be the *i-th* or the *$x^i$-component* of $f(x)$. The $f_i(x)$ themselves are called the *reduced components* of $f(x)$. However, if the component representation is written in the (equivalent) form

$$f(x) = \sum_{i=0}^{p-1} x^i f_i(x)^p$$

(where it should be considered that $f_i(x)$ has a different meaning to that in (1)), we shall understand by the reduced component the $f_i(x)$ occurring here. No misunderstandings will arise.

Now the *principle of comparison of components*, consisting of the identification of the $i$-th components ($i=0, \ldots, p-1$) of two equal polynomials, is based on the component representation; in most cases we immediately go over (without further reference) to the reduced components.

Component comparison is often better than coefficient comparison; its superiority originates in the finiteness of the number of the components, so that component comparison may be effective even in cases where coefficient comparison fails. In particular, it can be a useful tool for the solution of polynomial equations, leading to a *system of polynomial equations* in which the (reduced) components of the unknowns occur as new unknowns.

In the case of differential equations (over $\mathsf{F}[x]$) this method can prove to be particularly advantageous, because the differentiation of (1) yields the component representation

$$f'(x) = \sum_{i=0}^{p-1} ix^{i-1} f_i(x^p),$$

so that the behaviour of the components is essentially invariant inasmuch as they merely undergo a cyclic permutation and are supplied with simple numerical factors (elements of $\mathsf{F}_p$). Thus component comparison, applied to differential equations in $\mathsf{F}[x]$, serves as an "integration procedure".

NOTE 1. The concept of components is capable of generalization, of course, to power series as well as polynomials and for any natural number $k \geqq 2$ in place of $p$. We shall sometimes use this fact (without reference). In the case of such a generalization everything remains preserved except the invariance at differentiation; even this exception does not occur if $p|k$.

In $\mathsf{F}[x]$ and, more generally, in $\mathsf{F}((x))$ we shall also have to deal with differential equations which are integrable by a *quadrature*; by this we mean those differential equations which can be brought into the form

$$\left(\frac{f(x)}{g(x)}\right)' = 0 \qquad (f(x),\ g(x) \in \mathsf{F}[x];\ g(x) \neq 0). \tag{2}$$

The quadrature leads simply to

$$\frac{f(x)}{g(x)} = \frac{u(x^p)}{v(x^p)},$$

with arbitrary polynomials $u(x) \in \mathsf{F}[x]$, $v(x) \in \mathsf{F}[x] \setminus 0$.

Of course, (2) can be written, without any essential change, in the form

$$f'(x)g(x) - f(x)g'(x) = 0 \tag{3}$$

or

$$\frac{f'(x)}{f(x)} - \frac{g'(x)}{g(x)} = 0. \tag{4}$$

Some of the above remarks can be summarized for the purpose of later applications as follows:

PROPOSITION 10. *For polynomials $f(x), g(x) \in \mathsf{F}[x] \setminus 0$ the equivalence*

$$\begin{vmatrix} f(x) & g(x) \\ f'(x) & g'(x) \end{vmatrix} = 0 \Leftrightarrow \frac{f(x)}{g(x)} \in \mathsf{F}(x^p) \setminus 0$$

*holds.*

Differential equations of the form

$$\sum_{i=1}^{k} c_i \frac{f_i'(x)}{f_i(x)} = 0 \qquad (c_i \in \mathsf{F}_p \setminus 0;\ f_i(x) \in \mathsf{F}[x] \setminus 0) \tag{5}$$

will also occur. These equations, too, can be integrated by a quadrature. For this purpose let us choose the $c_i$ as elements of the set $\{1, \ldots, p-1\}$. Then (5) is equivalent to

$$(f_1(x)^{c_1} \ldots f_k(x)^{c_k})' = 0,$$

and thus the quadrature leads to the result

$$f_1(x)^{c_1} \ldots f_k(x)^{c_k} = u(x^p) \qquad (u(x) \in \mathsf{F}[x] \setminus 0). \tag{6}$$

(If, however, the $c_i$ lie in $\mathsf{F}$, then, in general, the quadrature cannot be applied to (5).)

NOTE 2. If in (5) the denominators were removed, e.g. by multiplication by their least common multiple, the integration would thereby become more difficult rather than easier.

NOTE 3. In the case of classical differential equations the quadrature can in general be regarded as the best (though of course only rarely practicable) method of integration and we are used to consider a differential equation as solved when the quadrature has been carried out. The state of affairs is different in the case of differential equations in $\mathsf{F}[x]$ as the latter are transformed by the quadrature into polynomial equations the solution of which can afford considerable, sometimes even insurmountable, difficulties. For this, as well as for other reasons the quadrature is far from always being the best method of integration in $\mathsf{F}[x]$. Even the $\lambda$-differential equation will in this respect reveal a remarkable "pathological" behaviour, about which we shall soon give more precise details.

## § 12. Reduction of the $\lambda$-differential equation to the $a$, $b$-polynomial equation

For the rest of this chapter we shall automatically leave the case $p=2$ out of consideration. This should be borne in mind by the reader.

We propose us to solve the $\lambda$-differential equation

$$(g(x)^2)' = x^q - x + \lambda g(x) \qquad \left(\lambda \in \mathsf{F},\ g(x) \in \mathsf{F}[x],\ g^\circ = \frac{q+1}{2}\right), \tag{1}$$

but it is merely a reduction in this task that we shall achieve here.

It should be noted that a "classical" differential equation of the form (1) in the case $\lambda \neq 0$ is not solvable by a quadrature.

Of all the results to be obtained for the present case, we stress the importance of the following:

1) The $\lambda$-differential equation (1) is solvable only for $\lambda = \pm 3$.

2) The solutions come from suitable solutions of the $a$, $b$-polynomial equation.

3) For at least $\frac{1}{2}\left(p - \left(\frac{2}{p}\right)\right)$ values of $\lambda$ (all lying in $\mathsf{F}_p$) the equation (1) is integrable by a quadrature.

1) and 3) apparently contradict one another. This contradiction is explained by the fact that (1) is changed by a quadrature into a polynomial equation. Although the latter is of a very simple form and it contains interesting information about (1), it cannot be solved for any $\lambda \neq \pm 3$. Another surprising fact is that the quadrature seems to be completely unsuited to the proof of the existence of solutions. All these are rather unusual conditions, hardly ever found in connection with classical differential equations. They bear witness that in $\mathsf{F}[x]$ quadrature is under certain circumstances a very unreliable procedure of integration.

Let us now make the following remarks. If in (1) we do not stipulate that $g(x)$ should be monic, no essential generalization is achieved. On the one hand, it is easily seen that then only $g(x)$ with the leading coefficient $-1$ can be added as "new" solutions and, on the other hand, these solutions arise from the "old" ones simply by multiplication by $-1$, since (1) remains invariant under the substitution of the pair $\lambda$, $g(x)$ by $-\lambda$, $-g(x)$. Thus we have proved that in (1) a restriction to monic solutions is not essential. We stick to (1) since we need only the monic solutions.

Before dealing with (1) we make the following observations.

We agree that in the sum operators of the form

$$\sum_{i_1+\cdots+i_m=k} \qquad (k=0, 1, \ldots,; \; m=1, 2, \ldots)$$

it is always only the values 0, 1, ... , that are admitted as summation variables (and therefore these sum operators lead to finite sums). We shall disregard such terms of a sum as are undefined (or we agree to take their values as zero).

For finitely or infinitely many elements $\alpha_0, \alpha_1, \ldots$ of a commutative ring with unity element 1 we introduce the useful and very short (nevertheless completely clear) notation

$$\overset{m}{\alpha}_k = \sum_{i_1+\cdots+i_m=k} \alpha_{i_1}\cdots\alpha_{i_m} \qquad (k=0, 1, \ldots; \; m=1, 2, \ldots).$$

The identity

$$\sum_{0\leqq k<\infty} \overset{m}{\alpha}_k x^k = \Big(\sum_{0\leqq k<\infty} \alpha_k x^k\Big)^m \qquad (m = 1, 2, \ldots)$$

holds for power series in all cases where both sides have a meaning. (In particular, this identity holds in $\mathsf{F}[[x]]$.)

PROPOSITION 11. *For every natural number $l$ the system of equations*

$$\overset{2}{\alpha}_k=0 \qquad (k=2, \ldots, l) \tag{2}$$

*for elements $\alpha_0, \ldots, \alpha_l$ of a field of characteristic $\neq 2$, subject to the "initial condition"*

$$\alpha_0 \neq 0 \tag{3}$$

*is equivalent to*

$$\alpha_k = \binom{\frac{1}{2}}{k}\left(\frac{2\alpha_1}{\alpha_0}\right)^k \alpha_0 \qquad (k = 0, \ldots, l). \tag{4}$$

(Thus, under the condition (3), (4) can be used as the complete solution of (2).)

To prove this, suppose that $\mathfrak{F}$ is the given field. We form the polynomial

$$f(x) = \sum_{k=0}^{l} \alpha_k x^k \qquad (\in \mathfrak{F}[x]).$$

By squaring we obtain

$$f(x)^2 = \sum_{k=0}^{2l} \overset{2}{\alpha}_k x^k.$$

Thus the system of equations (2) is equivalent to the congruence

$$f(x)^2 \equiv \overset{2}{\alpha}_0 + \overset{2}{\alpha}_1 x \qquad (\mathrm{mod}\ x^{l+1}).$$

As

$$\overset{2}{\alpha}_0 = \alpha_0^2, \quad \overset{2}{\alpha}_1 = 2\alpha_0\alpha_1$$

this congruence can be written in the form

$$\left(\frac{f(x)}{\alpha_0}\right)^2 \equiv 1 + \frac{2\alpha_1}{\alpha_0}x \qquad (\mathrm{mod}\ x^{l+1}).$$

If this congruence is thought of in the field $\mathfrak{F}$, then the root can be extracted. We obtain

$$\frac{f(x)}{\alpha_0} \equiv \sum_{0 \leqq k < \infty} \binom{\frac{1}{2}}{k} \left(\frac{2\alpha_1}{\alpha_0}\right)^k x^k \quad (\mathrm{mod}\ x^{l+1})$$

(where we have used the fact that on both sides the constant term is equal to 1). Multiplication by $\alpha_0$ gives the validity of Proposition 11 by comparing coefficients.

The discussion of the $\lambda$-differential equation (1) is facilitated if we temporarily introduce, as a new unknown, the polynomial

$$h(x) = x^{\frac{q+1}{2}} g\left(\frac{1}{x}\right) \qquad \left(h(x) \in \mathsf{F}[x],\ h^{\circ} \leqq \frac{q-1}{2},\ h(0) = 1\right), \tag{5}$$

reciprocal to $g(x)$. Thus

$$g(x) = x^{\frac{q+1}{2}} h\left(\frac{1}{x}\right). \tag{6}$$

After squaring and differentiating we obtain

$$(g(x)^2)' = x^q h\left(\frac{1}{x}\right)^2 - 2x^{q-1} h\left(\frac{1}{x}\right) h'\left(\frac{1}{x}\right).$$

After substituting this into (1) the transformation $x \to \dfrac{1}{x}$ gives

$$x^{-q} h(x)^2 - 2x^{-q+1} h(x) h'(x) = x^{-q} - x^{-1} + \lambda x^{-\frac{q+1}{2}} h(x),$$

whence, after multiplying by $x^q$, the equation

$$h(x)^2 - x(h(x)^2)' = 1 - x^{q-1} + \lambda x^{\frac{q-1}{2}} h(x) \tag{7}$$

is obtained. We shall call this equation, in view of (1), the *reciprocal $\lambda$-differential equation.* Of course, (1) and (7) are completely equivalent. Although (7) is in every respect more complicated than (1), as regards the form of these equations, it will soon be clear that (7) has the advantage that by (5) the constant term of $h(x)$ is equal to 1.

From equation (7) we reach the congruence

$$h(x)^2 - x(h(x)^2)' \equiv 1 \quad \left(\operatorname{mod} x^{\frac{q-1}{2}}\right).$$

Interpreting the latter in the field $\mathsf{F}((x))$, we can divide by $x^2$. Thus

$$\frac{h(x)^2 - x(h(x)^2)'}{x^2} \equiv \frac{1}{x^2} \quad \left(\operatorname{mod} x^{\frac{q-5}{2}}\right).$$

(The case $q=3$ does not constitute an exception.) Multiplying this by $-1$, we obtain

$$\left(\frac{h(x)^2}{x}\right)' \equiv \left(\frac{1}{x}\right)' \quad \left(\operatorname{mod} x^{\frac{q-5}{2}}\right)$$

and the quadrature leads to

$$\frac{h(x)^2}{x} \equiv \frac{1}{x} + u(x^p) \quad \left(\operatorname{mod} x^{\frac{q-3}{2}}\right)$$

with a suitable polynomial $u(x) \in \mathsf{F}[x]$. (An upper bound could be given for the degree $u^{\circ}$ of $u(x)$ but there is no need for doing this.) It follows that

$$h(x)^2 \equiv 1 + xu(x^p) \quad \left(\operatorname{mod} x^{\frac{q-1}{2}}\right).$$

By going over to the square roots

$$h(x) \equiv \sum_{0 \leq k < \infty} \binom{\frac{1}{2}}{k} x^k u(x^p)^k \quad \left(\operatorname{mod} x^{\frac{q-1}{2}}\right), \tag{8}$$

where (cf. (5)) $h(0)=1$ was used.

According to Lucas' theorem (Proposition 8), it follows from the $p$-adic expansion

$$\frac{1}{2} = \frac{p+1}{2} + \frac{p-1}{2} p(1+p+p^2+\cdots)$$

that in (8) the terms belonging to

$$k \equiv \frac{p+3}{2}, \ldots, p-1 \qquad (\mathrm{mod}\, p)$$

vanish. Since, moreover, because $h^\circ \leqq \frac{q+1}{2}$ (cf. (5)) at most two elements $\geqq \frac{q-1}{2}$ appear in the exponent set $\mathfrak{E}(h)$, viz. the numbers $\frac{q-1}{2}, \frac{q+1}{2}$, and these numbers are congruent mod $p$ to the numbers $\frac{p-1}{2}, \frac{p+1}{2}$, the component representation

$$h(x) = \sum_{k=0}^{\frac{p+1}{2}} x^k h_k(x^p) \qquad (h_k(x) \in \mathsf{F}[x]), \tag{9}$$

where vanishing components have been omitted, follows easily from (8). (Accordingly $h(x)$ is a lacunary polynomial.)

This result can be easily reformulated concerning $g(x)$. For, by (6) and (9) we have

$$g(x) = x^{\frac{q+1}{2}} \sum_{k=0}^{\frac{p+1}{2}} x^{-k} h_k\left(\frac{1}{x^p}\right).$$

As the right-hand side lies in $\mathsf{F}[x]$ and $\frac{q+1}{2} \equiv \frac{p+1}{2}$ (mod $p$), the equation

$$g(x) = \sum_{k=0}^{\frac{p+1}{2}} x^{\frac{p+1}{2}-k} g_k(x^p) \tag{10}$$

$$\left(g_k(x) \in \mathsf{F}[x];\ g_0(x) \underset{1}{\in} \mathsf{F}[x];\ g_k^\circ \leqq g_0^\circ = \frac{q-p}{2p}\right)$$

follows by a simple change of notation, where the conditions indicated in the brackets follow from those in (1). This equation expresses the component representation of $g(x)$, where the numbering of the components is a little unusual, $g_k(x^p)$ being the $\left(\frac{p+1}{2} - k\right)$-th component. We shall adhere this numbering because it will be suitable for our purposes at a later stage. (Vanishing components have, of course, also been omitted in (10).) Let us note that (10) is trivial for the case $p=3$, but this will cause no trouble.

Our next aim is the integration of (1) by component comparison. Having just obtained the component representation (10), we shall find this comparison particularly advantageous, as almost the half the components of $g(x)$ vanish because of (10). All the same, the task facing us is not an easy one.

For convenience, we complete the definition of the $g_k(x)$ that appear in (10) by defining

$$g_k(x) = 0 \qquad \left(k < 0 \quad \text{or} \quad k > \frac{p+1}{2}\right). \tag{11}$$

Then after squaring (10) becomes

$$g(x)^2 = \sum_{k=0}^{p+1} \sum_{i+j=k} x^{p+1-i-j} g_i(x^p) g_j(x^p),$$

i.e.

$$g(x)^2 = \sum_{k=0}^{p+1} (x^{p+1-k} \sum_{i+j=k} g_i(x^p) g_j(x^p)).$$

With the notation introduced before Proposition 11 we put

$$\overset{2}{g}_k(x) = \sum_{i+j=k} g_i(x) g_j(x) \qquad (k = 0, 1, \ldots). \tag{12}$$

Then the previous equation gives

$$g(x)^2 = \sum_{k=0}^{p+1} x^{p+1-k} \overset{2}{g}_k(x^p),$$

whence by differentiation

$$(g(x)^2)' = \sum_{k=0}^{p} (1-k) x^{p-k} \overset{2}{g}_k(x^p). \tag{13}$$

On the other hand, we transform (10) into

$$g(x) = \sum_{k=\frac{p-1}{2}}^{p} x^{p-k} g_{k-\frac{p-1}{2}}(x^p) \tag{14}$$

by the substitution $k \to k - \dfrac{p-1}{2}$.

Now let us substitute (13) and (14) into (1), and then compare the $x^{p-k}$-components on both sides with each other, doing this, however, (successively) only for

$$k = 1, \ldots, \frac{p+1}{2}, p-1, 0, \tag{15}$$

as this will be sufficient for our purposes. (More precisely, the comparison of the remaining components, as will transpire, would yield nothing new.) For $p=3$ the number 2 appears twice in (15) but this will not disturb us.

We obtain, first of all, the equations

$$(1-k)\overset{2}{g}_k(x^p) = \lambda g_{k-\frac{p-1}{2}}(x^p) \qquad \left(k = 1, \ldots, \frac{p+1}{2}\right)$$

(where the right-hand side vanishes, by (11), for $k = 1, \ldots, \frac{p-3}{2}$), then the equation

$$2\overset{2}{g}_{p-1}(x^p) = -1 + \lambda g_{\frac{p-1}{2}}(x^p)$$

and finally the equation

$$x^p \overset{2}{g}_0(x^p) + \overset{2}{g}_p(x^p) = x^q + \lambda g_{\frac{p-1}{2}}(x^p).$$

The substitution $x^p \to x$ gives the system of polynomial equations

$$(1-k)\overset{2}{g}_k(x) = \lambda g_{k-\frac{p-1}{2}}(x) \qquad \left(k = 1, \ldots, \frac{p+1}{2}\right), \tag{16}$$

$$2\overset{2}{g}_{p-1}(x) = -1 + \lambda g_{\frac{p-1}{2}}(x), \tag{17}$$

$$x\overset{2}{g}_0(x) + \overset{2}{g}_p(x) = x^r + \lambda g_{\frac{p+1}{2}}(x) \qquad \left(r = \frac{q}{p}\right). \tag{18}$$

Thus the component comparison has been carried out for (1). (We have introduced the abbreviated notation $r$ for $\frac{q}{p}$ and shall continue to use this, though not always. We note that the left-hand side of (16) vanishes for $k=1$; the right-hand side, too, vanishes, by (11), for $k=1$ if $p \neq 3$; for this reason (16) can be disregarded for $p \neq 3$, $k=1$.)

Our next task consists in the investigation of the system of polynomial equations (16), (17), (18) with $\frac{p+5}{2}$ unknowns; these are the reduced components $g_k(x)$ $\left(k = 0, \ldots, \frac{p+1}{2}\right)$ and the parameter $\lambda$. Among other things, we shall find by a rather complicated elimination procedure that all the $g_k(x)$ can be represented by the first two, so we write for simplicity:

$$a(x) = g_0(x), \quad b(x) = 2g_1(x). \tag{19}$$

(The numerical factor 2 in $(19_2)$ serves to simplify our final formulae.) However, we shall not make use of these notations until later.

For the time being we disregard the case $p=3$ but this case will not constitute an exception to our result.

We shall temporarily use the notation

$$G_k(x) = \binom{\frac{1}{2}}{k}\left(\frac{2g_1(x)}{g_0(x)}\right)^k g_0(x)(\in \mathsf{F}(x)) \tag{20}$$

$$(k=0, 1, \ldots;\ G_0(x)=g_0(x),\ G_1(x)=g_1(x)).$$

(The right-hand side has a meaning as $g_0(x)\neq 0$, by virtue of (10).)

From the system (16), (17), (18) we first consider only the equations (16), and even these provisionally only for $k = 2, \ldots, \frac{p-3}{2}$. As has already been pointed out, in these cases the right-hand side of (16) is equal to 0 and the numerical factor $1-k$ on the left-hand side is not divisible by $p$, so that it can be cancelled. Thus we have obtained the system of equations

$$\overset{2}{g}_k(x) = 0 \qquad \left(k = 2, \ldots, \frac{p-3}{2}\right).$$

(This system is empty for $p=5$, a fact which will cause no trouble.) Hence by (20) and Proposition 11 the validity of

$$g_k(x) = G_k(x) \qquad \left(k = 0, \ldots, \frac{p-3}{2}\right) \tag{21}$$

can be inferred. Let us note the secondary result, contained in (21), according to which the polynomial quotients (20) must be integer for every solution $g(x)$ of (1), if $k \leqq \frac{p-3}{2}$; we shall return to this subject later in another form.

We now avail ourselves of the two remaining equations (16), belonging to the values $k = \frac{p-1}{2}, \frac{p+1}{2}$; these equations run as follows:

$$\frac{3}{2}\overset{2}{g}_{\frac{p-1}{2}}(x) = \lambda g_0(x), \qquad \frac{1}{2}\overset{2}{g}_{\frac{p+1}{2}}(x) = \lambda g_1(x).$$

Now applying (21) on the right-hand sides, we obtain the equations

$$\frac{3}{2}\overset{2}{g}_{\frac{p-1}{2}}(x) = \lambda G_0(x), \qquad \frac{1}{2}\overset{2}{g}_{\frac{p+1}{2}}(x) = \lambda G_1(x). \tag{22}$$

We introduce the notation

$$\overset{2}{G}_k(x) = \sum_{i+j=k} G_i(x)G_j(x) \qquad (k = 0, 1, \ldots), \tag{23}$$

similar to (12). Then by (20), and Proposition 11

$$\overset{2}{G}_k(x)=0 \qquad (k=2,3,\ldots).$$

$\left(\text{For } k=2,\ldots,\frac{p-3}{2}\right.$ this relation has already been noted above, before (21).)

From the cases $k=\frac{p-1}{2},\frac{p+1}{2}$, (22) can be replaced by

$$\frac{3}{2}\left(\overset{2}{g}_{\frac{p-1}{2}}(x)-\overset{2}{G}_{\frac{p-1}{2}}(x)\right)=\lambda G_0(x), \qquad \frac{1}{2}\left(\overset{2}{g}_{\frac{p-1}{2}}(x)-\overset{2}{G}_{\frac{p+1}{2}}(x)\right)=\lambda G_1(x). \tag{24}$$

The values of both expressions in brackets on the left-hand sides are simply obtained from (12) and (23), applying (21) (as several terms drop out),

$$2G_0(x)\left(g_{\frac{p-1}{2}}(x)-G_{\frac{p-1}{2}}(x)\right)$$

and

$$2G_0(x)\left(g_{\frac{p+1}{2}}(x)-G_{\frac{p+1}{2}}(x)\right)+2G_1(x)\left(g_{\frac{p-1}{2}}(x)-G_{\frac{p-1}{2}}(x)\right),$$

respectively. Thus dividing by $G_0(x)$ $(=g_0(x)\neq 0)$ the equations (24) transform into the equations

$$3\left(g_{\frac{p-1}{2}}(x)-G_{\frac{p-1}{2}}(x)\right)=\lambda$$

and

$$g_{\frac{p+1}{2}}(x)-G_{\frac{p+1}{2}}(x)+\frac{G_1(x)}{G_0(x)}\left(g_{\frac{p-1}{2}}(x)-G_{\frac{p-1}{2}}(x)\right)=\lambda\frac{G_1(x)}{G_0(x)}.$$

As $p\neq 3$, the first equation is divisible by 3, whence

$$g_{\frac{p-1}{2}}(x)-G_{\frac{p-1}{2}}(x)=\frac{\lambda}{3}. \tag{25}$$

By substitution the second equation takes the simpler form

$$g_{\frac{p+1}{2}}(x)-G_{\frac{p+1}{2}}(x)=\frac{2\lambda}{3}\frac{G_1(x)}{G_0(x)}. \tag{26}$$

In order to summarize the results obtained so far, we now introduce the notations (19) into (20), getting

$$G_k(x)=\binom{\frac{1}{2}}{k}\left(\frac{b(x)}{a(x)}\right)^k a(x) \qquad (k=0,1,\ldots).$$

Substituting this in (21), (25) and (26) we arrive at

$$g_k(x) = \binom{\frac{1}{2}}{k}\left(\frac{b(x)}{a(x)}\right)^k a(x) = \begin{cases} 0 & \left(k = 0, \ldots, \frac{p-3}{2}\right), \\ \frac{\lambda}{3} & \left(k = \frac{p-1}{2}\right), \\ \frac{\lambda}{3}\,\frac{b(x)}{a(x)} & \left(k = \frac{p+1}{2}\right). \end{cases} \tag{27}$$

From this result we shall very soon need the special cases $k = \frac{p-3}{2}, \frac{p-1}{2}, \frac{p+1}{2}$. In order to put these cases into the simplest possible form, we use the fact that

$$\binom{\frac{1}{2}}{k} \equiv \binom{\frac{1}{2}}{\frac{p+1}{2}-k} \pmod p \quad \left(k = 0, \ldots, \frac{p+1}{2}\right).$$

For the values of $k$ just mentioned the right-hand side of this congruence is equal to $-\frac{1}{8}, \frac{1}{2}$ or 1, respectively. After a few transformations we thus obtain from (27) the following three equations

$$g_{\frac{p-3}{2}}(x) = -\frac{1}{8}\left(\frac{b(x)}{a(x)}\right)^{\frac{p-3}{2}} a(x), \tag{28}$$

$$g_{\frac{p-1}{2}}(x) = \frac{\lambda}{3} + \frac{1}{2}\left(\frac{b(x)}{a(x)}\right)^{\frac{p-1}{2}} a(x), \tag{29}$$

$$g_{\frac{p+1}{2}}(x) = \frac{b(x)}{a(x)}\left(\frac{\lambda}{3} + \left(\frac{b(x)}{a(x)}\right)^{\frac{p-1}{2}} a(x)\right). \tag{30}$$

We now consider equation (17). By (11) and (12), the polynomial $\overset{2}{g}_{p-1}(x)$ is equal to

$$g_{\frac{p-1}{2}}(x)^2 + 2g_{\frac{p-3}{2}}(x)g_{\frac{p+1}{2}}(x),$$

and therefore (17) turns after slight transformations into

$$g_{\frac{p-1}{2}}(x)\left(2g_{\frac{p-1}{2}}(x) - \lambda\right) + 4g_{\frac{p-3}{2}}(x)g_{\frac{p+1}{2}}(x) = -1. \tag{31}$$

On the other hand, (28), (29) and (30) can be written, temporarily using the abbreviation

$$\Omega = \left(\frac{b(x)}{a(x)}\right)^{\frac{p-1}{2}} a(x),$$

in the form

$$g_{\frac{p-3}{2}}(x) = -\frac{1}{8}\,\Omega\,\frac{a(x)}{b(x)},$$

$$g_{\frac{p-1}{2}}(x) = \frac{\lambda}{3} + \frac{1}{2}\,\Omega,$$

$$g_{\frac{p+1}{2}}(x) = \frac{b(x)}{a(x)}\left(\frac{\lambda}{3} + \Omega\right),$$

with the exception of the case $b(x)=0$ which will be settled subsequently to give the same result. Thus $\left(\text{as } \dfrac{a(x)}{b(x)}\dfrac{b(x)}{a(x)}=1\right)$ substitution into (31) gives the equation

$$\left(\frac{\lambda}{3} + \frac{1}{2}\,\Omega\right)\left(-\frac{\lambda}{3} + \Omega\right) - \frac{1}{2}\,\Omega\left(\frac{\lambda}{3} + \Omega\right) = -1,$$

which, when multiplied out, gives

$$-\left(\frac{\lambda}{3}\right)^2 = -1. \tag{32}$$

The case $b(x)=0$, which we have so far not considered is no exception, as then, by (28), (29) and (30), we have simply

$$g_{\frac{p-3}{2}}(x)=0, \quad g_{\frac{p-1}{2}}(x)=\frac{\lambda}{3}, \quad g_{\frac{p+1}{2}}(x)=0,$$

whence, by (31), equation (32) follows also in this case. After solving this, we write it in the final form

$$\lambda = 3\sigma \qquad (\sigma = \pm 1). \tag{33}$$

This important partial result shows that the $\lambda$-differential equation can be solved for only two values of the parameter, at most which was promised earlier. (We shall obtain the same result for the case $p=3$ neglected so far which will soon be dealt with. In this case, however, (33) is to be read as $\lambda=0$, so that there is then only one "admissible" value of $\lambda$.)

Finally, we consider also the only remaining equation, (18). By (11) and (12) this equation becomes simply

$$xg_0(x)^2+2g_{\frac{p-1}{2}}(x)g_{\frac{p+1}{2}}(x) = x^r+\lambda g_{\frac{p+1}{2}}(x) \qquad \left(r=\frac{q}{p}\right),$$

which can be replaced by

$$xg_0(x)^2+g_{\frac{p+1}{2}}(x)\left(-\lambda+2g_{\frac{p-1}{2}}(x)\right) = x^r.$$

Hence by $(19_1)$, (29), and (30)

$$xa(x)^2+\frac{b(x)}{a(x)}\left(\frac{\lambda}{3}+\left(\frac{b(x)}{a(x)}\right)^{\frac{p-1}{2}}a(x)\right)\left(-\frac{\lambda}{3}+\left(\frac{b(x)}{a(x)}\right)^{\frac{p-1}{2}}a(x)\right) = x^r.$$

After multiplying out and substituting from (32) we get

$$xa(x)^2+\frac{b(x)}{a(x)}\left(-1+\left(\frac{b(x)}{a(x)}\right)^{p-1}a(x)^2\right) = x^r.$$

Multiplying out again and then multiplying through by $a(x)^{p-2}$, after rearranging the terms we obtain

$$xa(x)^p+b(x)^p = x^r a(x)^{p-2}+a(x)^{p-3}b(x) \tag{34}$$

$$\left(r=\frac{q}{p};\ a(x)\underset{1}{\in}\mathsf{F}[x];\ b(x)\in\mathsf{F}[x];\ b^{\circ}\leqq a^{\circ}=\frac{r-1}{2}\right).$$

$\Bigl($The conditions in brackets follow from (10), (19) and $\frac{q-p}{2p}=\frac{r-1}{2}.\Bigr)$ The reader will note that (34) is identical with the $a$, $b$-polynomial equation. (We shall obtain similar results for $p=3$.)

We now calculate $g(x)$ from (10), (27) and (33):

$$g(x) = \sum_{k=0}^{\frac{p+1}{2}} x^{\frac{p+1}{2}-k}\binom{\frac{1}{2}}{k}\left(\frac{b(x^p)}{a(x^p)}\right)^k a(x^p)+\sigma\left(x+\frac{b(x^p)}{a(x^p)}\right).$$

We can write $\frac{p+1}{2}$ instead of $\frac{1}{2}$ in the binomial coefficients; thus we have (for the moment for $p\neq 3$)

$$g(x) = a(x^p)\left(x+\frac{b(x^p)}{a(x^p)}\right)^{\frac{p+1}{2}}+\sigma\left(x+\frac{b(x^p)}{a(x^p)}\right). \tag{35}$$

The remaining case $p=3$ can easily be disposed of. Now for $k=1$ (because of $\frac{p-1}{2}=1$) (16) amounts merely to

$$0=\lambda g_0(x).$$

Since the second factor does not vanish, this implies $\lambda=0$. We see that (33) holds for this case too.

The equations (17) and (18) now take (because $\lambda=0$) the form

$$\overset{2}{g}_2(x)=1, \quad x\overset{2}{g}_0(x)+\overset{2}{g}_3(x)=x^r.$$

By virtue of (11) and (12), these equations state that

$$2g_0(x)g_2(x)+g_1(x)^2 = 1, \quad xg_0(x)^2+2g_1(x)g_2(x) = x^r.$$

After substituting from (19) we have (partly because $p=3$)

$$2a(x)g_2(x)+b(x)^2 = 1, \quad xa(x)^2+b(x)g_2(x) = x^r.$$

It is obvious that this system is equivalent to the following one:

$$xa(x)^3+b(x)^3 = x^r a(x)+b(x), \quad g_2(x) = \frac{b(x)^2-1}{a(x)}. \tag{36}$$

The first of these equations is just the case $p=3$ of (34). (Of course, there is no change in the conditions in the brackets.) By the second equation we obtain from (10) and (19)

$$g(x) = x^2a(x^3)+2xb(x^3)+\frac{b(x^3)^2-1}{a(x^3)},$$

i.e.

$$g(x) = a(x^3)\left(x+\frac{b(x^3)}{a(x^3)}\right)^2-\frac{1}{a(x^3)}. \tag{37}$$

We have obtained similar results to those for $p\neq 3$, the only difference being that the last equation is not the case $p=3$ for (35). This exception can easily be removed in the following way. We introduce a "new" $b(x)$ by means of the substitution

$$b(x)\to b(x)-\sigma,$$

where $\sigma$ may be equal to 1 or $-1$. As $\sigma\in \mathsf{F}_3$, $(36_1)$ remains invariant, whereas (37) becomes

$$g(x)\left(= a(x^3)\left(x+\frac{b(x^3)-\sigma}{a(x^3)}\right)^2-\frac{1}{a(x^3)}\right) = a(x^3)\left(x+\frac{b(x^3)}{a(x^3)}\right)^2+\sigma\left(x+\frac{b(x^3)}{a(x^3)}\right).$$

This equation is actually the case $p=3$ for (35). (Of course, for $p=3$ $(19_2)$

no longer holds for the "new" $b(x)$, but this does not matter because we shall no longer use (19) in what follows.)

We now state

THEOREM 6. *The $\lambda$-differential equation*

$$(g(x)^2)' = x^q - x + \lambda g(x) \qquad \left(\lambda \in \mathsf{F};\ g(x) \underset{1}{\in} \mathsf{F}[x];\ g^\circ = \frac{q+1}{2}\right) \tag{38}$$

*is solvable if and only if*

$$\lambda = 3\sigma \qquad (\sigma = \pm 1); \tag{39}$$

*moreover, in this case all its solutions are obtained by substituting the solutions $a(x)$, $b(x)$ of the a, b-polynomial equation*

$$xa(x)^p + b(x)^p = x^r a(x)^{p-2} + a(x)^{p-3} b(x) \tag{40}$$

$$\left(r = \frac{q}{p};\ a(x) \underset{1}{\in} \mathsf{F}[x];\ b(x) \in \mathsf{F}[x];\ b^\circ \leqq a^\circ = \frac{r-1}{2}\right)$$

*into*

$$g(x) = a(x^p)\left(x + \frac{b(x^p)}{a(x^p)}\right)^{\frac{p+1}{2}} + \sigma\left(x + \frac{b(x^p)}{a(x^p)}\right), \tag{41}$$

*retaining only those of the resulting expressions $g(x)$ that fall in* $\mathsf{F}[x]$.

(It will be seen from the proof that the remaining $g(x)$, given by (41), namely those falling in $\mathsf{F}(x) \setminus \mathsf{F}[x]$, also satisfy equation (38), but this is irrelevant to our purpose.)

The "only if" part of the theorem has already been proved. Further, we have seen (cf. (34) and (35)) that all solutions of (38) (also in the case $p=3$) are supplied by (40) and (41). We still have to prove the "if" part and that in case (39) the $g(x) \in \mathsf{F}[x]$, satisfying (40) and (41), are solutions of (38).

We first prove the latter assertion. For this purpose we shall use the abbreviation

$$\Omega = x + \frac{b(x^p)}{a(x^p)}. \tag{42}$$

Let us divide (40) by $a(x)^{p-2}$. From the substitution $x \to x^p$ we obviously obtain

$$a(x^p)^2 \Omega^p = x^q - x + \Omega. \tag{43}$$

On the other hand, (41) becomes

$$g(x) = a(x^p)\Omega^{\frac{p+1}{2}} + \sigma\Omega. \tag{44}$$

Since by (42) the derivative of $\Omega$ is equal to 1 we have

$$g'(x) = \frac{1}{2} a(x^p)\,\Omega^{\frac{p-1}{2}} + \sigma.$$

Thus

$$(g(x)^2)' = g(x)2g'(x) = a(x^p)^2\,\Omega^p + 3\sigma a(x^p)\,\Omega^{\frac{p+1}{2}} + 2\sigma^2\,\Omega.$$

On the right-hand side we substitute (40), which gives (partly by virtue of $\sigma^2 = 1$)

$$(g(x)^2)' = x^q - x + 3\sigma\left(a(x^p)\,\Omega^{\frac{p+1}{2}} + \sigma\Omega\right).$$

Bearing (39) and (44) in mind, this shows that equation (38) is satisfied. By (42), (44) and $g(x) \in \mathsf{F}[x]$, the conditions in (38) follow from those in (40). Thus we have proved the assertion.

Finally, in order to prove the "if" part of the theorem, we assume the validity of (39). We note that the $a$, $b$-polynomial equation (40) has the solution

$$a(x) = x^{\frac{r-1}{2}}, \quad b(x) = 0,$$

which we shall call its *trivial solution.* The substitution of this solution into (41) leads to the polynomial

$$g(x) = x^{\frac{q+1}{2}} + \sigma x,$$

which is a solution of the $\lambda$-differential equation (38), as a consequence of the part of Theorem 6 already proved (this can also easily be verified directly). This accomplishes the proof of Theorem 6.

NOTE 1. With this theorem we have achieved the promised reduction of the $\lambda$-differential equation to the $a$, $b$-polynomial equation. This is really only an "incomplete" reduction, as one has not merely to substitute the solutions of (40) into the solution formula (41) of (38) but one still has then to select the integer $g(x)$, i.e. those falling in $\mathsf{F}[x]$. This task constitutes a divisibility problem which we shall briefly call here the *$\lambda$-divisibility problem.* Thus, to find all solutions of the $\lambda$-differential equation, one has first to determine those of the $a$, $b$-polynomial equation and then to solve the $\lambda$-divisibility problem. This is a lengthy program, both problems being very difficult.

We now apply Theorem 6 to the special differential equation

$$(g(x)^2)' = x^q - x + \left(\varkappa + \frac{2}{\varkappa}\right) g(x) \qquad \left(\varkappa \in \mathsf{F} \setminus 0;\ g(x) \underset{1}{\in} \mathsf{F}[x];\ g^{\circ} = \frac{q+1}{2}\right), \tag{45}$$

which we are most interested in; thus we extend this application to the marginal case of Problem II as well. (39) shows that (45) is solvable only for the $\varkappa$ with

$$\varkappa+\frac{2}{\varkappa}=3\sigma \qquad (\sigma=\pm 1); \tag{46}$$

these are the four parameter values

$$\varkappa=\sigma,\, 2\sigma \qquad (\sigma=\pm 1). \tag{47}$$

(For $p=3$ there are only the two values $\varkappa=\sigma=\pm 1$.)

Now we make use of Proposition 9 by which the solutions of Problem II, for the marginal case, are

$$f(x)=x^q-x+\varkappa g(x), \tag{48}$$

where $\varkappa$ and $g(x)$ satisfy (45). (Only those $f(x)$ which are fully reducible are to be retained.) Before substituting $g(x)$ from (41), we first transform (48). In order to do this, we again apply the abbreviation (42). (48) then by (43) becomes

$$f(x)=a(x^p)^2\,\Omega^p-\Omega+\varkappa g(x). \tag{49}$$

We also use the form (44) of (41). After the substitution into (49) we get

$$f(x)=a(x^p)^2\,\Omega^p+\varkappa a(x^p)\,\Omega^{\frac{p+1}{2}}+(\varkappa\sigma-1)\,\Omega.$$

Corresponding to the two possibilities (47), we deduce that

$$f(x)=a(x^p)\,\Omega^{\frac{p+1}{2}}\left(a(x^p)\,\Omega^{\frac{p-1}{2}}+\sigma\right) \quad \text{or} \quad f(x)=\Omega\left(a(x^p)\,\Omega^{\frac{p-1}{2}}+\sigma\right)^2,$$

which can be summarized as

$$f(x)=\Omega\left(a(x^p)\,\Omega^{\frac{p-1}{2}}+\sigma\right)\left(a(x^p)\,\Omega^{\frac{p-1}{2}}+\sigma\tau\right) \qquad (\sigma=\pm 1,\ \tau=0,1),$$

with the help of a further parameter $\tau$. Finally, substitution of (42) gives

PROPOSITION 12. *All solutions of Problem II, for the marginal case, are given by substituting the solutions of the a, b-polynomial equation*

$$xa(x)^p+b(x)^p=x^r a(x)^{p-2}+a(x)^{p-3}b(x) \tag{50}$$

$$\left(r=\frac{q}{p};\ a(x)\underset{1}{\in}\mathsf{F}[x],\ b(x)\in\mathsf{F}[x];\ b^\circ\leqq a^\circ=\frac{r-1}{2}\right)$$

*into*

$$f(x)=\left(x+\frac{b(x^p)}{a(x^p)}\right)\left(a(x^p)\left(x+\frac{b(x^p)}{a(x^p)}\right)^{\frac{p-1}{2}}+\sigma\right)\left(a(x^p)\left(x+\frac{b(x^p)}{a(x^p)}\right)^{\frac{p-1}{2}}+\sigma\tau\right) \tag{51}$$

$$(\sigma=\pm 1;\ \tau=0,1),$$

*retaining from these polynomial quotients $f(x)$ only those ones which lie in* $\mathsf{F}[x]$ *and are fully reducible.*

(Formula (51) will, after we have solved the $a, b$-polynomial equation, be replaced by another, much simpler one. It could be shown that, for $p=3$, $\tau=0$ would be sufficient in (51), but we shall return—only later and in another form—to this subject.)

NOTE 2. It is seen from the above that no trouble could have been avoided if, instead of the $\lambda$-differential equation (38), we had restricted ourselves from the very outset to the differential equation (45).

NOTE 3. As has been shown by L. Megyesi, the necessity of the solvability condition $\lambda=3\sigma$ $(\sigma=\pm 1)$, formulated in Theorem 6, can be proved directly as follows. For a solution $g(x)$ of the $\lambda$-differential equation both equations

$$2g'(x)^2+2g(x)g''(x) = -1+\lambda g'(x), \quad \sigma g'(x)g''(x)+2g(x)g''(x) = \lambda g''(x)$$

can be deduced by differentiating once or twice, respectively. On the other hand, as has been pointed out, the divisibility $g(x)|x^q-x$ holds, and therefore $g(x)$ has no multiple zeros. It should be also remembered that $g(x)\neq 0$, as $g^{\circ} = \dfrac{p+1}{2}$. Hence the existence of a $\xi\in\bar{\mathsf{F}}$ follows with $g(\xi)=0$, $g''(\xi)\neq 0$. (Even $\xi\in\mathsf{F}$ holds here.) The substitution $x\to\xi$ leads (partly after dividing by $g''(\xi)$) to the equations

$$-2g'(\xi)^2+\lambda g'(\xi) = 1, \quad 6g'(\xi) = \lambda.$$

We actually obtain

$$4g'(\xi)^2 = 1, \quad g'(\xi) = \pm\frac{1}{2}, \quad \lambda = \pm 3.$$

This easy proof provides no substantial simplification of the proof of Theorem 6.

## § 13. The quadrature of the $\lambda$-differential equation

Our investigations would be incomplete if we did not discuss the quadrature of the $\lambda$-differential equation

$$(g(x)^2)' = x^q-x+\lambda g(x) \qquad \left(\lambda\in\mathsf{F},\ g(x)\underset{1}{\in}\mathsf{F}[x],\ g^{\circ} = \frac{q-1}{2}\right), \tag{1}$$

which we shall do here, briefly. Initially we take no notice of the fact that, by Theorem 6, (1) is solvable only for $\lambda=\pm 3$, admitting for $\lambda$ all values from $\mathsf{F}$.

Since

$$g(x)|x^q-x \tag{2}$$

follows from (1), it seems to be appropriate to introduce

$$h(x) = \frac{x^q - x}{g(x)} \tag{3}$$

as a new unknown. To do this, we write (3) in the form

$$g(x)h(x) = x^q - x. \tag{4}$$

After differentiating we have

$$g'(x)h(x) + g(x)h'(x) + 1 = 0. \tag{5}$$

If the left-hand side of (4) is substituted for $x^q - x$ in (1), then after dividing by $g(x)$ we get

$$2g'(x) = h(x) + \lambda. \tag{6}$$

Elimination of $g'(x)$ from (5) and (6) gives

$$2g(x)h'(x) + h(x)^2 + \lambda h(x) + 2 = 0.$$

Multiplying by $h(x)$ and once again taking account of (4) we obtain the equation

$$2(x^q - x)h'(x) + h(x)^3 + \lambda h(x)^2 + 2h(x) = 0, \tag{7}$$

which may be called the *inverse λ-differential equation.* After its derivation, (7) is equivalent to (1).

It is easy to find such $\lambda \in \mathsf{F}$ for which equation (7) is integrable by a quadrature. To this end we put (7) in the form

$$\frac{2h'(x)}{h(x)^3 + \lambda h(x)^2 + 2h(x)} = \frac{1}{x - x^q}. \tag{8}$$

On the right-hand side the numerator is the derivative of the denominator. Hence it follows that (8) and thus also (7) and (1) are integrable by a quadrature if all the roots of the quadratic equation

$$\tau^2 + \lambda\tau + 2 = 0, \tag{9}$$

belonging to the denominator of the left-hand side of (8), lie in the prime field $\mathsf{F}_p$ and are distinct. Of course, the appropriate $\lambda$ themselves lie also in $\mathsf{F}_p$, and it is well-known that there are $\frac{1}{2}\left(p - \left(\frac{2}{p}\right)\right)$ of them. As (1) is solvable only for

$$\lambda = 3\sigma \qquad (\sigma = \pm 1), \tag{10}$$

it follows from this that for (1) "solvability by quadrature" is possible even without "solvability".

We shall return to this paradox later, but first we shall consider thoroughly case (10) (where solvability holds in both senses). Now (9) has the solutions $\tau = -\sigma, -2\sigma$. Accordingly, after taking partial fractions in (8) it becomes

$$\frac{h'(x)}{h(x)} - 2\frac{h'(x)}{h(x)+\sigma} + \frac{h'(x)}{h(x)+2\sigma} = \frac{1}{x-x^q},$$

and consequently the quadrature yields

$$\frac{h(x)(h(x)+2\sigma)}{(x^q-x)(h(x)+\sigma)^2} \in \mathsf{F}(x^p) \setminus 0.$$

Substituting from (3), we can replace this with

$$\frac{x^q - x + 2\sigma g(x)}{(x^q - x + \sigma g(x))^2} \in \mathsf{F}(x^p) \setminus 0.$$

This result can be summarized in the following

Supplement (to Theorem 6). *For $\lambda = 3\sigma$ $(\sigma = \pm 1)$ a polynomial $g(x) \in \mathsf{F}[x]$ of degree $g^\circ = \dfrac{q+1}{2}$ is a solution of the $\lambda$-differential equation if and only if the quotient of the polynomials*

$$x^q - x + 2\sigma g(x), \quad (x^q - x + \sigma g(x))^2$$

*is a p-th power in $\mathsf{F}(x)$.*

Although this supplement is not totally devoid of interest, we could not use it for the determination of the solutions of $g(x)$. The latter will be therefore determined from Theorem 6, disregarding the supplement. (On the strength of the supplement we should not even get any nearer to the connection between the $\lambda$-differential equation and the $a$, $b$-polynomial equation formulated in Theorem 6.)

Of course, the situation does not improve for other cases integrable by quadrature, which elucidates the above paradox. There is no need to give a more detailed explanation of this question.

Note 1. At the beginning of our investigations when we still believed in the absolute superiority of integration by quadrature, it was our luck that we had then made no attempt to solve the $\lambda$-differential equation by quadrature. If we had done that too early, we might well have given up all hope of success, but we obtained the above result on quadrature only when it could no longer influence us because we already had Theorem 6.

Note 2. It was only after the completion of this manuscript that P. Turán drew my attention to a classical differential equation (J. Surányi and P.

Turán [15], p. 76) which, although integrable by quadrature, using a multiplying factor, still has the property that those of its solutions that are polynomials of given formal degree (which are required for dealing with a special interpolation problem "with accessory conditions") had to be determined without quadrature, by a roundabout approach.

## § 14. The reciprocal *a*, *b*-polynomial equation and the *a*, *b*-power series equation

In order to facilitate the solution of the $a, b$-polynomial equation

$$xa(x)^p + b(x)^p = x^r a(x)^{p-2} + a(x)^{p-3} b(x) \tag{1}$$

$$\left(r = \frac{q}{p};\; a(x) \underset{1}{\in} \mathsf{F}[x],\; b(x) \in \mathsf{F}[x];\; b^\circ \leqq a^\circ = \frac{r-1}{2}\right),$$

let us use on it the transformation $x \to \dfrac{1}{x}$ by which we shall understand that, instead of $a(x)$ and $b(x)$, the reciprocal polynomials

$$\bar{a}(x) = x^{\frac{r-1}{2}} a\left(\frac{1}{x}\right), \quad \bar{b}(x) = x^{\frac{r-1}{2}} b\left(\frac{1}{x}\right) \tag{2}$$

are introduced as new unknowns.

From (2) the similar formulae

$$a(x) = x^{\frac{r-1}{2}} \bar{a}\left(\frac{1}{x}\right), \quad b(x) = x^{\frac{r-1}{2}} \bar{b}\left(\frac{1}{x}\right) \tag{3}$$

can be inferred. By substituting these into (1) and transforming by $x \to \dfrac{1}{x}$ we obtain

$$x^{-1-p\frac{r-1}{2}} \bar{a}(x)^p + x^{-p\frac{r-1}{2}} \bar{b}(x)^p =$$

$$= x^{-r-(p-2)\frac{r-1}{2}} \bar{a}(x)^{p-2} + x^{-(p-2)\frac{r-1}{2}} \bar{a}(x)^{p-3} \bar{b}(x),$$

i.e.

$$\bar{a}(x)^p + x\bar{b}(x)^p = \bar{a}(x)^{p-2} + x^r \bar{a}(x)^{p-3} \bar{b}(x) \tag{4}$$

$$\left(r = \frac{q}{p};\; \bar{a}(x), \bar{b}(x) \in \mathsf{F}[x];\; \bar{a}(0) = 1;\; \bar{a}^\circ, \bar{b}^\circ \leqq \frac{r-1}{2}\right),$$

where the conditions, indicated in brackets, follow from those in (1) and from (2). (By virtue of $\bar{a}(0) = 1$, (4) is more suitable than (1) for the determination of the solutions.)

We shall call (4) the polynomial equation reciprocal to (1), or simply the *reciprocal a, b-polynomial equation*. In this equation we shall henceforth use the simpler notation $a(x)$ and $b(x)$, instead of $\bar{a}(x)$ and $\bar{b}(x)$, which will not lead to any misunderstanding.

It is clear that (1) and (4) are completely equivalent to one another and therefore it suffices to investigate (4) instead of (1) (the results, nevertheless, will be used for (1)).

Our next important step consists in introducing the extension $\mathsf{F}[x] \to \mathsf{F}[[x]]$ of the solution domain. This means that we determine, more generally than in (4), all solutions of the equation

$$a(x)^p + xb(x)^p = a(x)^{p-2} + x^r a(x)^{p-3} b(x) \tag{5}$$

$$\left(r = \frac{q}{p};\ a(x), b(x) \in \mathsf{F}[[x]];\ a(0) = 1\right),$$

which we shall briefly call the *a, b-power series equation.*

Our next program consists of the following. We first determine the solutions of the *a, b*-power series equation. This will be done in § 15 by a uniformization process. Then the solutions lying in $\mathsf{F}[x]$ remain to be selected. This will give us the solutions of (4) and consequently also those of (1). This program, however, will turn out to be very complicated and we shall discuss it in §§ 16, 17 and 18.

## § 15. Uniformization of the *a, b*-power series equation

In order to solve the *a, b*-power series equation

$$a(x)^p + xb(x)^p = a(x)^{p-2} + x^r a(x)^{p-2} b(x) \tag{1}$$

$$\left(r = \frac{q}{p};\ a(x), b(x) \in \mathsf{F}[[x]];\ a(0) = 1\right)$$

we perform the *birational transformation*

$$a(x) = \mathfrak{q}(x)^{\frac{p-1}{2}}, \quad b(x) = \mathfrak{p}(x)\mathfrak{q}(x)^{\frac{p-3}{2}} \tag{2}$$

$$(\mathfrak{p}(x), \mathfrak{q}(x) \in \mathsf{F}[[x]];\ \mathfrak{q}(0) = 1)$$

by which the power series $\mathfrak{p}(x)$ and $\mathfrak{q}(x)$ are introduced as new unknowns. This is possible, as the latter can be calculated uniquely from (2) in the form

$$\mathfrak{p}(x) = a(x)^{-\frac{p-3}{p-1}} b(x), \quad \mathfrak{q}(x) = a(x)^{\frac{2}{p-1}}. \tag{3}$$

By substituting (2) into (1) and dividing by

$$\mathfrak{q}(x)^{p\frac{p-3}{2}} \tag{4}$$

we get the (fortunately simple) equation

$$\mathfrak{q}(x)^p + x\mathfrak{p}(x)^p = \mathfrak{q}(x) + x^r\mathfrak{p}(x). \tag{5}$$

Now from this equation $\mathfrak{q}(x)$ can be expressed in terms of $\mathfrak{p}(x)$. This will give the promised uniformization of (1). For this purpose we raise (5) to the $p^i$-th power:

$$\mathfrak{q}(x)^{p^{i+1}} + x^{p^i}\mathfrak{p}(x)^{p^{i+1}} = \mathfrak{q}(x)^{p^i} + x^{rp^i}\mathfrak{p}(x)^{p^i} \qquad (i = 0, 1, \ldots).$$

By addition for the initial values $i=0, \ldots, k$ we obtain the equation

$$\mathfrak{q}(x)^{p^{k+1}} + \sum_{i=0}^{k} x^{p^i}\mathfrak{p}(x)^{p^{i+1}} = \mathfrak{q}(x) + \sum_{i=0}^{k} x^{rp^i}\mathfrak{p}(x)^{p^i} \qquad (k = 0, 1, \ldots),$$

after cancelling the terms that appear on both sides. For the first term we have (as a consequence of $\mathfrak{q}(0)=1$)

$$\lim_{k\to\infty} \mathfrak{q}(x)^{p^k} = 1,$$

and thus, taking the limit, we have

$$1 + \sum_{0\leqq i<\infty} x^{p^i}\mathfrak{p}(x)^{p^{i+1}} = \mathfrak{q}(x) + \sum_{0\leqq i<\infty} x^{rp^i}\mathfrak{p}(x)^{p^i},$$

were we have used the fact that both infinite sums are convergent. Hence

$$\mathfrak{q}(x) = 1 + \sum_{0\leqq i<\infty} (x\mathfrak{p}(x)^p - x^r\mathfrak{p}(x))^{p^i}. \tag{6}$$

We shall show that, conversely, for every power series $\mathfrak{p}(x) \in \mathsf{F}[[x]]$ the formulae (2) and (6) provide a solution $a(x)$, $b(x)$ of (1).

Equation (6) implies that

$$\mathfrak{q}(x)^p = 1 + \sum_{0\leqq i<\infty} (x\mathfrak{p}(x)^p - x^r\mathfrak{p}(x))^{p^{i+1}}.$$

Hence, and again from (6), equation (5) follows by subtraction. If we multiply (5) by the power (4), we arrive at equation (1) as a result of (2). Since it follows from (2) and from (6) that the conditions $a(x), b(x) \in \mathsf{F}[[x]]$ are also satisfied, our assertion is proved.

For these reasons the (integer) power series $\mathfrak{p}(x)$ is called a *uniformizing parameter* for (1) and the result summarized in

THEOREM 7. *All solutions of the a, b-power series equation*

$$a(x)^p + xb(x)^p = a(x)^{p-2} + x^r a(x)^{p-3} b(x) \tag{7}$$

$$\left(r = \frac{q}{p};\ a(x), b(x) \in \mathsf{F}[[x]];\ a(0) = 1\right)$$

*are given by the solution formulae*

$$a(x) = \mathfrak{q}(x)^{\frac{p-1}{2}}, \quad b(x) = \mathfrak{p}(x)\mathfrak{q}(x)^{\frac{p-3}{2}} \tag{8}$$

$$\left(\mathfrak{q}(x) = 1 + \sum_{0 \leqq i < \infty} (x\mathfrak{p}(x)^p - x^r \mathfrak{p}(x))^{p^i}\right),$$

*where the uniformizing parameter* $\mathfrak{p}(x)$ *has to run through all power series from* $\mathsf{F}[[x]]$.

NOTE. It is trivial that all *unity units* $\mathfrak{e}(x) \in \mathsf{F}[[x]]$, i.e. the units $\mathfrak{e}(x)$ with $\mathfrak{e}(0) = 1$ can be uniquely represented in the form

$$\mathfrak{e}(x) = 1 + \sum_{0 \leqq i < \infty} \mathfrak{f}(x)^{p^i} \quad (\mathfrak{f}(x) \in x\mathsf{F}[[x]]),$$

where

$$\mathfrak{f}(x) = \mathfrak{e}(x) - \mathfrak{e}(x)^p$$

must hold. Therefore $\mathfrak{f}(x)$ may be called the *kernel* of $\mathfrak{e}(x)$. We see that the essential content of formula $(8_3)$ consists merely in the fact that the kernel of $\mathfrak{q}(x)$ is of the special form

$$x\mathfrak{p}(x)^p - x^r \mathfrak{p}(x)$$

(cf. (5)). This fact will be used extensively in § 17.

## § 16. A proposition on polynomials

Our task in this section is to prove, in preparation for the solution of the $a, b$-polynomial equation, the difficult

PROPOSITION 13. *For every* $s = 1, p, p^2, \ldots$ *the assumptions*

$$\mathfrak{u}(x), \mathfrak{a}(x), \mathfrak{b}_1(x), \ldots, \mathfrak{b}_{\frac{p-1}{2}}(x) \in \bar{\mathsf{F}}[x], \tag{1}$$

$$\mathfrak{u}(0) = 1, \quad \mathfrak{a}(0) = 1, \tag{2}$$

$$\mathfrak{u}^\circ \leqq s-1, \quad \mathfrak{a}^\circ \leqq \frac{s-1}{p-1}, \quad \mathfrak{b}_l^\circ \leqq \frac{s-1}{2} \quad \left(l = 1, \ldots, \frac{p-1}{2}\right) \tag{3}$$

*and*

$$\mathfrak{u}(x)^l \mathfrak{a}(x)^{\frac{p-1}{2}-l} \equiv \mathfrak{b}_l(x) \pmod{x^s} \quad \left(l = 1, \ldots, \frac{p-1}{2}\right) \tag{4}$$

*imply that*

$$\mathfrak{u}^\circ \leqq \frac{s-1}{p-1}. \tag{5}$$

The reader will notice that here the algebraic closure $\bar{\mathsf{F}}$ has been chosen as the fundamental field. This generality is necessary for our purpose. We shall return to this in § 17. (The notations used in (1) were designed with later applications in mind.)

Proposition 13 is trivially true for $p=3$ as (for $l=1$) the equation $\mathfrak{u}(x)= =\mathfrak{b}_1(x)$ then follows from the congruence (4), using ($3_1$) and ($3_3$), and therefore (5) is identical to ($3_3$). For the equally trivial case $s=1$ (5) follows from ($3_1$). After disregarding both trivial cases we still have to consider the case

$$s \geqq p \geqq 5. \tag{6}$$

The proof will consist essentially in a divisibility investigation in which we will consider the multiplicities $e_l$, defined by

$$\mathfrak{a}(x)^{e_l} \| \mathfrak{b}_l(x) \qquad \left(l = 1, \ldots, \frac{p-1}{2}\right), \tag{7}$$

with which $\mathfrak{a}(x)$ appears in the $\mathfrak{b}_l(x)$ as a divisor.

From (2) and (4) we infer that

$$\mathfrak{b}_l(0) = 1 \tag{8}$$

holds, whence $\mathfrak{b}_l(x) \neq 0$. Consequently, the definition (7) has a meaning if and only if the condition

$$\mathfrak{a}^\circ \geqq 1 \tag{9}$$

is satisfied (and then all multiplicities $e_l$ are unambiguously determined by (7)).

For the case when (9) does not hold, i.e. $\mathfrak{a}^\circ \leqq 0$, we start with a simple proof of Proposition 13.

Because of ($2_2$) we now have $\mathfrak{a}(x)=1$, and consequently (4) simply becomes

$$\mathfrak{u}(x)^l \equiv \mathfrak{b}_l(x) \pmod{x^s} \quad \left(l = 1, \ldots, \frac{p-1}{2}\right).$$

For $l=1$ this entails

$$\mathfrak{u}^\circ \leqq \frac{s-1}{2},$$

by $(3_1)$ and $(3_3)$. We assume more generally for a natural number $l < \frac{p-1}{2}$ the truth of the inequality

$$\mathfrak{u}^{\circ} \leqq \frac{s-1}{2l}$$

already proved, and prove that it then also holds for $l+1$ in place of $l$, whence (5) follows.

By the induction hypothesis we have

$$(\mathfrak{u}(x)^{l+1})^{\circ} = (l+1)\mathfrak{u}^{\circ} \leqq (l+1)\frac{s-1}{2l} \leqq s-1.$$

From this and from $(3_3)$ the preceding congruence, with $l+1$ in place of $l$, becomes the equation

$$\mathfrak{u}(x)^{l+1} = \mathfrak{b}_{l+1}(x).$$

Again in view of $(3_3)$ this actually implies that

$$\mathfrak{u}^{\circ} = \frac{1}{l+1}\mathfrak{b}_{l+1}^{\circ} \leqq \frac{s-1}{2(l+1)}.$$

Thus we have proved Proposition 13 for the case $\mathfrak{a}^{\circ} \leqq 0$.

So now let us assume (9) to be true, whence it follows, as has already been pointed out, that the multiplicities $e_l$, defined by (7), exist. We may therefore write the polynomials $\mathfrak{b}_l(x)$ in the form

$$\mathfrak{b}_l(x) = \mathfrak{a}(x)^{e_l}\mathfrak{r}_l(x)$$

so that (partly from $(2_2)$ and (8))

$$\mathfrak{r}_l(x) \in \bar{F}[x], \quad \mathfrak{a}(x) \nmid \mathfrak{r}_l(x), \quad \mathfrak{r}_l(0) = 1 \quad \left(l = 1, \ldots, \frac{p-1}{2}\right). \tag{10}$$

Putting this in (4) we have

$$\mathfrak{u}(x)^l \mathfrak{a}(x)^{\frac{p-1}{2}-l} \equiv \mathfrak{a}(x)^{e_l}\mathfrak{r}_l(x) \pmod{x^s} \quad \left(l = 1, \ldots, \frac{p-1}{2}\right) \tag{11}$$

and, further, the truth of

$$e_l \mathfrak{a}^{\circ} + \mathfrak{r}_l^{\circ} \leqq \frac{s-1}{2} \qquad \left(l = 1, \ldots, \frac{p-1}{2}\right) \tag{12}$$

can be inferred from $(3_3)$. (The $\mathfrak{b}_l(x)$ will no longer be used.)

We rewrite (11) with $k$ in place of $l$:

$$\mathfrak{u}(x)^k \mathfrak{a}(x)^{\frac{p-1}{2}-k} \equiv \mathfrak{a}(x)^{e_k} \mathfrak{r}_k(x) \pmod{x^s} \quad \left(k = 1, \ldots, \frac{p-1}{2}\right) \tag{13}$$

and eliminate $\mathfrak{u}(x)$ from (11) and (13). For this purpose we raise (11) and (13) to the powers $k$ and $l$, respectively. Then $\mathfrak{u}(x)$ will in both congruences have the same exponent $kl$ and therefore, as a result of the desired elimination, we obtain from them the congruence

$$\begin{vmatrix} \mathfrak{a}(x)^{\frac{p-1}{2}k-kl} & \mathfrak{a}(x)^{\frac{p-1}{2}l-kl} \\ \mathfrak{a}(x)^{ke_l} \mathfrak{r}_l(x)^k & \mathfrak{a}(x)^{le_k} \mathfrak{r}_k(x)^l \end{vmatrix} \equiv 0 \pmod{x^s} \quad \left(k, l = 1, \ldots, \frac{p-1}{2}\right),$$

from which only the case $k < l$ will be used in what follows. As $(\mathfrak{a}(x), x) = 1$ (cf. $(2_2)$), divisions by a power of $\mathfrak{a}(x)$ are allowed here. Thus the first row of the determinant can be replaced by

$$1, \mathfrak{a}(x)^{\frac{p-1}{2}(l-k)}$$

and consequently, after expanding (and rearranging the terms),

$$\mathfrak{a}(x)^{le_k} \mathfrak{r}_k(x)^l \equiv \mathfrak{a}(x)^{ke_l + \frac{p-1}{2}(l-k)} \mathfrak{r}_l(x)^k \pmod{x^s} \quad \left(1 \leqq k < l \leqq \frac{p-1}{2}\right). \tag{14}$$

We consider the special case $l = kt$ ($t = 2, 3, \ldots$) of (14). Then only exponents divisible by $k$ occur in (14), and consequently we can take the $k$-th root, so that

$$\mathfrak{a}(x)^{te_k} \mathfrak{r}_k(x)^t = \mathfrak{a}(x)^{e_{kt} + \frac{p-1}{2}(t-1)} \mathfrak{r}_{kt}(x) \pmod{x^s} \tag{15}$$

$$\left(k = 1, 2, \ldots; \ t = 2, 3, \ldots; \ kt \leqq \frac{p-1}{2}\right).$$

(Here the addition of no constant factor was necessary, as by $(2_2)$ and $(10_3)$ both sides of (15) have the same constant term 1.)

From (15), too, we consider only the special case $t = 2 \left(k \leqq \frac{p-1}{4}\right)$. Then the degrees of the left- and right-hand sides are respectively equal to

$$2(e_k \mathfrak{a}^\circ + \mathfrak{r}_k^\circ) \quad \text{and} \quad \frac{p-1}{2} \mathfrak{a}^\circ + (e_{2k} \mathfrak{a}^\circ + \mathfrak{r}_{2k}^\circ).$$

So it follows from (12) and partly from ($3_2$) that both degrees are $\leqq s-1$; accordingly the congruences (15) for this case become the equations

$$\mathfrak{a}(x)^{2e_k}\mathfrak{r}_k(x)^2 = \mathfrak{a}(x)^{e_{2k}+\frac{p-1}{2}}\mathfrak{r}_{2k}(x) \qquad \left(1 \leqq k \leqq \left[\frac{p-1}{4}\right]\right). \tag{16}$$

This and ($10_2$) obviously imply that

$$2e_k \leqq e_{2k}+\frac{p-1}{2} \leqq 2e_k+1 \qquad \left(1 \leqq k \leqq \left[\frac{p-1}{4}\right]\right), \tag{17}$$

whence

$$e_k = \left[\frac{e_{2k}+\frac{p-1}{2}}{2}\right] \geqq \left[\frac{p-1}{4}\right] \qquad \left(1 \leqq k \leqq \left[\frac{p-1}{4}\right]\right). \tag{18}$$

Thus we certainly have

$$e_1 \geqq \frac{p-3}{4}. \tag{19}$$

Henceforth we distinguish between two cases.

We first consider the case

$$e_1 \geqq \frac{p-3}{2}, \tag{20}$$

for which we prove that

$$e_k \geqq \frac{p-1}{2}-k, \quad \mathfrak{u}^\circ \leqq \frac{1}{k}\left(\frac{s-1}{2}-\left(\frac{p-1}{2}-k\right)\mathfrak{a}^\circ\right) \qquad \left(k = 1, \ldots, \frac{p-1}{2}\right) \tag{21}$$

and note that for $k = \frac{p-1}{2}$ ($21_2$) is identical with (5), i.e. it is equivalent to Proposition 13, still to be proved.

We first carry out the proof of (21) for $k=1$. Now, by (20), only ($21_2$) remains to be proved. We consider the congruence (11) for $l=1$; it is seen from (20) that this congruence can be divided by the power of $\mathfrak{a}(x)$ occurring on the left-hand side. This gives the congruence

$$\mathfrak{u}(x) \equiv \mathfrak{a}(x)^{e_1-\frac{p-3}{2}}\mathfrak{r}_1(x) \pmod{x^s}.$$

By reason of ($3_1$) and (12), both sides are of degree $\leqq s-1$, and thus even the equation

$$\mathfrak{u}(x) = \mathfrak{a}(x)^{e_1-\frac{p-3}{2}}\mathfrak{r}_1(x),$$

holds. Hence

$$\mathfrak{u}^\circ = \left(e_1 - \frac{p-3}{2}\right)\mathfrak{a}^\circ + \mathfrak{r}_1^\circ .$$

Consequently, again by (12),

$$\mathfrak{u}^\circ \leqq \frac{s-1}{2} - \frac{p-3}{2}\mathfrak{a}^\circ .$$

This inequality agrees with the case $k=1$ of $(21_2)$. Thus assertion (21) is proved for this case.

Now let $k \geqq 2$. For $k-1$ we assume the validity of (21). We consider the congruence (13) and show that its left-hand side is of degree $\leqq s-1$, i.e. that

$$k\mathfrak{u}^\circ + \left(\frac{p-1}{2} - k\right)\mathfrak{a}^\circ \leqq s-1.$$

As by the induction hypothesis $(21_2)$ holds for $k-1$ (in place of $k$), it suffices to prove that

$$\frac{k}{k-1}\left(\frac{s-1}{2} - \left(\frac{p-1}{2} - (k-1)\right)\mathfrak{a}^\circ\right) + \left(\frac{p-1}{2} - k\right)\mathfrak{a}^\circ \leqq s-1.$$

Multiplication by $k-1$ turns this inequality into

$$k\left(\frac{s-1}{2} - \left(\frac{p-1}{2} - (k-1)\right)\mathfrak{a}^\circ\right) + (k-1)\left(\frac{p-1}{2} - k\right)\mathfrak{a}^\circ \leqq (k-1)(s-1),$$

the validity of which, in fact, follows trivially from $k \geqq 2$ and (9) as the left-hand side is equal to

$$\frac{k}{2}(s-1) - \frac{p-1}{2}\mathfrak{a}^\circ .$$

This proves that the left-hand side of (13) is of degree $\leqq s-1$. For the right-hand side the same holds, by (12). Therefore (13) implies the equation

$$\mathfrak{u}(x)^k \mathfrak{a}(x)^{\frac{p-1}{2} - k} = \mathfrak{a}(x)^{e_k}\mathfrak{r}_k(x).$$

In view of $(10_2)$ the power of $\mathfrak{a}(x)$ on the right-hand side must be at least as high as that on the left-hand side, and this proves $(21_1)$. Furthermore, by comparing degrees we get

$$k\mathfrak{u}^\circ + \left(\frac{p-1}{2} - k\right)\mathfrak{a}^\circ = e_k\mathfrak{a}^\circ + \mathfrak{r}_k^\circ .$$

From (12), the right-hand side is $\leqq \frac{s-1}{2}$, and we have thus also proved ($21_2$). We have thereby proved Proposition 13 for the case (20).

We still have to consider the case

$$e_1 \leqq \frac{p-5}{2}. \tag{22}$$

It will turn out, however, that this case cannot arise. For, we shall prove the assertion

$$(k-1)\frac{p-3}{2k} \leqq e_1 \leqq \frac{p-5}{2}, \quad \mathfrak{r}_1(x)^k = \mathfrak{a}(x)^{e_k+(k-1)\frac{p-1}{2}-ke_1}\mathfrak{r}_k(x) \tag{23}$$

$$\left(k = 1, \ldots, \frac{p-1}{2}\right),$$

of which, however, ($23_1$) is not true for $k = \frac{p-1}{2}$, as can easily be seen.

For $k=1$ ($23_1$) follows from (22); and, further, ($23_2$) is satisfied identically. Henceforth let $k \geqq 2$ and let us assume the truth of (23) for $k-1$ (in place of $k$). We apply (14) with the pair $1, k$ instead of $k, l$:

$$\mathfrak{a}(x)^{ke_1}\mathfrak{r}_1(x)^k \equiv \mathfrak{a}(x)^{e_k+\frac{p-1}{2}(k-1)}\mathfrak{r}_k(x) \pmod{x^s}.$$

This congruence can be divided by the power

$$\mathfrak{a}(x)^{\min\left(ke_1, \frac{p-1}{2}(k-1)\right)},$$

whence

$$\mathfrak{a}(x)^{\max\left(0, ke_1-\frac{p-1}{2}(k-1)\right)}\mathfrak{r}_1(x)^k \equiv \mathfrak{a}(x)^{e_k+\max\left(0, \frac{p-1}{2}(k-1)-ke_1\right)}\mathfrak{r}_k(x) \pmod{x^s}. \tag{24}$$

Let $G_1$ and $G_2$ denote the degrees of the left- and right-hand sides respectively: we shall show that

$$G_1 \leqq s-1, \quad G_2 \leqq s-1. \tag{25}$$

This implies the equality of the two sides of (24), i.e. the validity of ($23_2$) and, in consequence of ($10_2$) (and $e_k \geqq 0$), the inequality

$$(k-1)\frac{p-1}{2} - ke_1 \leqq k-1,$$

i.e.

$$(k-1)\frac{p-3}{2} \leqq ke_1;$$

together with (22), this last inequality gives $(23_1)$, which will give the proof of Proposition 13.

We have

$$G_1 = k\mathfrak{r}_1^{\circ} + \max\left(0,\, ke_1 - \frac{p-1}{2}(k-1)\right)\mathfrak{a}^{\circ}.$$

On the other hand, we note that (12) and $(23_2)$ imply the inequality

$$k\mathfrak{r}_1^{\circ} \leqq \frac{s-1}{2} + \left(\frac{p-1}{2}(k-1) - ke_1\right)\mathfrak{a}^{\circ}.$$

Thus, by the induction hypothesis, the latter is valid for $k-1$ in place of $k$. Therefore

$$(k-1)\mathfrak{r}_1^{\circ} \leqq \frac{s-1}{2} + \left(\frac{p-1}{2}(k-2) - (k-1)e_1\right)\mathfrak{a}^{\circ}$$

holds, from which we easily deduce that

$$(k-1)G_1 \leqq k\frac{s-1}{2} + \left(\frac{p-1}{2}(k^2-2k) - (k^2-k)e_1\right)\mathfrak{a}^{\circ} +$$

$$+\max\left(0,\, (k^2-k)e_1 - \frac{p-1}{2}(k^2-2k+1)\right)\mathfrak{a}^{\circ},$$

i.e.

$$(k-1)G_1 \leqq k\frac{s-1}{2} + \max\left(\frac{p-1}{2}(k^2-2k) - (k^2-k)e_1,\, -\frac{p-1}{2}\right)\mathfrak{a}^{\circ}.$$

As, on the other hand, $(23_1)$ is, by the induction hypothesis, valid for $k-1$ (in place of $k$),

$$(k-2)\frac{p-3}{k(k-1)} \leqq e_1$$

holds; consequently

$$-(k^2-k)e_1 \leqq -\frac{p-3}{2}(k^2-2k). \tag{26}$$

Thus from the above inequality

$$(k-1)G_1 \leqq k\frac{s-1}{2} + \max\left(k^2-2k,\, -\frac{p-1}{2}\right)\mathfrak{a}^{\circ},$$

i.e.

$$(k-1)G_1 \leqq k\frac{s-1}{2} + (k^2-2k)\mathfrak{a}^{\circ}$$

can be inferred. Hence and from $(3_2)$ it follows that

$$(k-1)G_1 \leqq \left(\frac{k}{2}+\frac{k^2-2k}{p-1}\right)(s-1),$$

i.e.

$$2(k-1)(p-1)G_1 \leqq \big(k(p-1)+2k^2-4k\big)(s-1). \tag{27}$$

Thus, in order to prove $(25_1)$, it is sufficient to show that

$$2(k-1)(p-1) \geqq k(p-1)+2k^2-4k$$

is true. This inequality is equivalent to

$$(k-2)(p-1) \geqq 2k^2-4k,$$

and therefore it is correct, since $p-1 \geqq 2k$. This completes the proof of $(25_1)$.

For $G_2$ we have

$$G_2 = \left(e_k+\max\left(0, \frac{p-1}{2}(k-1)-ke_1\right)\mathfrak{a}^\circ+\mathfrak{r}_k^\circ,\right.$$

so, by (12),

$$G_2 \leqq \frac{s-1}{2}+\max\left(0, \frac{p-1}{2}(k-1)-ke_1\right)\mathfrak{a}^\circ.$$

Hence, and from (26), we infer that

$$(k-1)G_2 \leqq (k-1)\frac{s-1}{2}+\max\left(0, \frac{p-1}{2}(k^2-2k+1)-\frac{p-3}{2}(k^2-2k)\right)\mathfrak{a}^\circ,$$

i.e.

$$(k-1)G_2 \leqq (k-1)\frac{s-1}{2}+\max\left(0, k^2-2k+\frac{p-1}{2}\right)\mathfrak{a}^\circ,$$

i.e.

$$(k-1)G_2 \leqq (k-1)\frac{s-1}{2}+\left(k^2-2k+\frac{p-1}{2}\right)\mathfrak{a}^\circ.$$

By $(3_2)$ it follows that

$$2(k-1)(p-1)G_2 \leqq \big((k-1)(p-1)+(2k^2-4k+p-1)\big)(s-1),$$

i.e.

$$2(k-1)(p-1)G_2 \leqq \big(k(p-1)+2k^2-4k\big)(s-1).$$

This inequality is of the form (27) (with $G_2$ in place of $G_1$) and therefore, by the preceding deduction, we have proved $(25_2)$ and thus also Proposition 13.

## § 17. A proposition on power series

We shall now try to use the last proposition, to prove a proposition for certain power series in $\bar{\mathsf{F}}[[x]]$ which will enable us in § 18 to solve the *a, b*-polynomial equation with the help of Theorem 7. (We draw the reader's attention to the fact that once again the algebraic closure $\bar{\mathsf{F}}$ has been taken as the fundamental field, the necessity of which we shall justify more fully below.)

We must make a few preliminary remarks, partly also with later aims in view, some of which refer to arbitrary power series $\mathfrak{a}(x) \in \bar{\mathsf{F}}((x))$ and integers $k$, $l$.

For power series $\mathfrak{a}(x) \in \bar{\mathsf{F}}(x))$ and integers $k$ we denote by $(\mathfrak{a}(x))_{x^k}$ the *residue* of $\mathfrak{a}(x)$ mod $x^k$ and define this as the sum of the terms of $\mathfrak{a}(x)$ possessing a degree $<k$. (It may be seen that this is a generalization of a symbol already introduced for polynomials.)

We define the *coefficient operator* $\langle\ \rangle_k$ to mean the coefficient of $x^k$ in $\mathfrak{a}(x)$. Clearly

$$\langle \alpha x^l \mathfrak{a}(x^p)\rangle_{pk+l} = \alpha \langle \mathfrak{a}(x)\rangle_k \qquad (\alpha \in \bar{\mathsf{F}}).$$

Therefore for $\alpha \in \bar{\mathsf{F}} \setminus 0$ the inclusion

$$\langle \alpha x^l \mathfrak{a}(x^p)\rangle_{pk+l} = 0 \Rightarrow \langle \mathfrak{a}(x)\rangle_k = 0$$

holds which we shall use without further reference. ($\Leftrightarrow$ would hold here in place of $\Rightarrow$ but this is of no relevance for us.)

By $\mathfrak{a}^p(x)$ we shall understand the power series arising from $\mathfrak{a}(x)$ by applying the automorphism $\xi \to \xi^p$ of $\bar{\mathsf{F}}$ on the coefficients. We have the rule

$$\mathfrak{a}(x)^p = \mathfrak{a}^p(x^p).$$

Consequently for $\alpha \neq 0$ the other form

$$\langle \alpha x^l \mathfrak{a}(x)^p\rangle_{pk+l} = 0 \Rightarrow \langle \mathfrak{a}(x)\rangle_k = 0$$

of the former inclusion is also valid, and is likewise used without reference in what follows.

Clearly, the unique representation

$$\mathfrak{a}(x) = \sum_{i=0}^{p-1} x^i \mathfrak{a}_i(x^p) \qquad (\mathfrak{a}_i(x) \in \bar{\mathsf{F}}(x))$$

holds with suitable $\mathfrak{a}_i(x)$; we continue to call this the *component representation* of $\mathfrak{a}(x)$ even in this more general sense. The inclusion

$$\langle \mathfrak{a}(x) \rangle_{pk+l} = 0 \Rightarrow \langle \mathfrak{a}_l(x) \rangle_k = 0$$

follows trivially from this for each $l=0, \ldots, p-1$.

For $k \leqq l$ we understand by the (formal) equation

$$\langle \mathfrak{a}(x) \rangle_{k,l} = 0$$

(to the left-hand side of which we attach no independent meaning) that all

$$\langle \mathfrak{a}(x) \rangle_i = 0 \qquad (k \leqq i \leqq l)$$

are valid. (Thus the equation $\langle \mathfrak{a}(x) \rangle_{k,k}=0$ means the same as $\langle \mathfrak{a}(x) \rangle_k=0$.)

A polynomial from $\mathsf{F}[x]$ will be called a *p-polynomial* for short if for each positive element $k$ of its exponent set there exists an $i=0, 1, \ldots$ with

$$p^i \leqq k \leqq \frac{p^{i+1}-1}{p-1} \, (= p^i + \cdots + p + 1),$$

i.e. its exponent set lies in the set

$$\{0, 1;\; p, p+1;\; p^2, \ldots, p^2+p+1;\; p^3, \ldots,\; p^3+p^2+p+1;\; \ldots\}.$$

(Accordingly, the $p$-linear polynomials are special $p$-polynomials.)

PROPOSITION 14. *With a fixed natural number $c$ let us write*

$$r = p^c, \quad s = p^{c-1}. \tag{1}$$

*If*

$$\mathfrak{p}(x) \in \bar{\mathsf{F}}[[x]], \quad \mathfrak{p}(0) = 1, \tag{2}$$

*then, putting*

$$\mathfrak{v}(x) = \sum_{0 \leqq i < \infty} (x\mathfrak{p}(x)^p)^{p^i} \left(= x\mathfrak{p}(x)^p + x^p \mathfrak{p}(x)^{p^2} + x^{p^2} \mathfrak{p}(x)^{p^3} + \ldots\right) \tag{3}$$

*it follows from the assumption*

$$\left\langle (1+\mathfrak{v}(x))^{\frac{p-1}{2}} \right\rangle_{\frac{r+1}{2}, r-1} = 0 \tag{4}$$

*that the residue $(\mathfrak{p}(x))_{x^s}$ is a p-polynomial.*

NOTE. We have introduced $\bar{\mathsf{F}}$ as the algebraic closure of $\mathsf{F}$, but we have already pointed out that $\bar{\mathsf{F}}$ is also the algebraic closure of $\mathsf{F}_p$; thus Proposition 14 does not depend on our fundamental field $\mathsf{F}$. On the other hand, we shall apply it only for the case

$$\mathfrak{p}(x) \in \mathsf{F}[[x]], \quad c = n-1 \quad \left(r = \frac{q}{p}, \; s = \frac{q}{p^2}, \; q \geqq p^2\right).$$

This case is, in two senses, less general than Proposition 14 but the full generality of Proposition 14 is necessary in order to carry out the following induction in the proof (this fact has been alluded to already); we have not managed to find another proof.

For $c=1$ Proposition 14 is trivially true, as then $s=1$. Henceforth let $c \geqq 2$ and let us suppose, moreover, that the assertion is true for $c-1$ in place of $c$.

We introduce (as a continuation of (1)) the notation

$$t=p^{c-2}. \tag{5}$$

($t$ is an integer because $c \geqq 2$.)

To prove the assertion, we split it into the following two parts:

**a** The residue $(\mathfrak{p}(x))_{x^t}$ is a $p$-polynomial.

**b** The inequality $(\mathfrak{p}(x))^{\circ}_{x^s} \leqq \frac{s-1}{p-1} \left(= t+\frac{t}{p}+\cdots+p+1\right)$ holds.

In order to prove **a**, let us write (4) in the form

$$\left\langle (1+\mathfrak{v}(x))^{\frac{p-1}{2}} \right\rangle_k = 0 \qquad \left(\frac{r+1}{2} \leqq k \leqq r-1\right). \tag{6}$$

We now consider only the part relating to $p|k$. Since (from (1)) the multiples of $p$ lying between $\frac{r+1}{2}$ and $r-1$ can be written in the form

$$pk \qquad \left(\frac{s+1}{2} \leqq k \leqq s-1\right),$$

from (6) we have

$$\left\langle (1+\mathfrak{v}(x))^{\frac{p-1}{2}} \right\rangle_{pk} = 0 \qquad \left(\frac{s+1}{2} \leqq k \leqq s-1\right). \tag{7}$$

On the other hand, from (3) the identity

$$1+\mathfrak{v}(x) = x\mathfrak{p}(x)^p+1+\mathfrak{v}(x)^p \tag{8}$$

follows. Thus, from the binomial theorem, we have the component representation

$$(1+\mathfrak{v}(x))^{\frac{p-1}{2}} = \sum_{i=0}^{\frac{p-1}{2}} \binom{-\frac{1}{2}}{i} x^i \mathfrak{p}(x)^{pi}(1+\mathfrak{v}(x)^p)^{\frac{p-1}{2}-i} \tag{9}$$

which will also be used later. (Here we could write $-\frac{1}{2}$ instead of $\frac{p-1}{2}$ in the binomial coefficient.) When we substitute (9) into (7) we must retain only the 0-th component (i.e. the term belonging to $i=0$) and therefore (7) becomes

$$\left\langle (1+\mathfrak{v}(x)^p)^{\frac{p-1}{2}} \right\rangle_{pk} = 0 \qquad \left(\frac{s+1}{2} \leqq k \leqq s-1\right).$$

Thus

$$\left\langle (1+\mathfrak{v}(x))^{\frac{p-1}{2}} \right\rangle_{k} = 0 \qquad \left(\frac{s+1}{2} \leqq k \leqq s-1\right).$$

This means that (4) is satisfied with $s$ in place of $r$, and thus it follows from the induction hypothesis (cf. (1) and (5)) that the residue $(\mathfrak{p}(x))_{x^t}$ is a $p$-polynomial. We have thus proved **a**.

To prove **b**, we start with the remark that for every $p$-polynomial $f(x)$ $f(x)^p$ and $xf(x)^p$ are clearly also $p$-polynomials. It follows from repeated application of this and from **a** that $x(\mathfrak{p}(x))^p_{x^t}$ and the powers

$$\left(x(\mathfrak{p}(x))^p_{x^t}\right)^{p^i} \qquad (i=0, 1, \ldots)$$

are $p$-polynomials. As, however,

$$(\mathfrak{p}(x))^p_{x^k} = (\mathfrak{p}(x)^p)_{x^{pk}} \qquad (k=0, 1, \ldots)$$

obviously holds whence (because $pt=s$) it follows that,

$$x(\mathfrak{p}(x))^p_{x^t} = (x\mathfrak{p}(x)^p)_{x^{s+1}},$$

we conclude that all powers

$$(x\mathfrak{p}(x)^p)^{p^i}_{x^{s+1}} \qquad (i=0, 1, \ldots)$$

are $p$-polynomials. Since the latter can be written as

$$\left((x\mathfrak{p}(x)^p)^{p^i}\right)_{x^{p^i(s+1)}} \qquad (i=0, 1, \ldots),$$

*a fortriori* all residues

$$\left((x\mathfrak{p}(x)^p)^{p^i}\right)_{x^s} \qquad (i=0, 1, \ldots)$$

are $p$-polynomials (as $p^i(s+1) > s-1$). On account of (3) it follows from this by adding that the residue

$$(\mathfrak{v}(x))_{x^s}$$

is likewise a $p$-polynomial. As this polynomial is, however, of degree $\leqq s-1$, i.e. all elements of its exponent set are less than $s$, these elements are, according to the definition of the $p$-polynomials, not greater than $\frac{s-1}{p-1}$. Consequently

$$(\mathfrak{v}(x))^{\circ}_{x^s} \leqq \frac{s-1}{p-1}. \tag{10}$$

(This partial result is already very similar to the assertion to be proved, inasmuch as it means that (10) holds for $\mathfrak{p}(x)$ in place of $\mathfrak{v}(x)$.)

In what follows we shall often write the non-negative integer numbers in the *canonical representation*

$$pk+l \qquad (k=0, 1, \dots;\ l=0, \dots, p-1) \tag{11}$$

(dependent from p). Using these, (6) can be written in the form

$$\left\langle (1+\mathfrak{v}(x))^{\frac{p-1}{2}} \right\rangle_{pk+l} = 0 \qquad \left(\frac{r+1}{2} \leqq pk+l \leqq r-1\right). \tag{12}$$

The component representation (9) shows that (12) is trivially satisfied in the cases $l=\frac{p+1}{2}, \dots, p-1$ and that the latter may therefore be left out of consideration. We do the same for the case $l=0$, considered above (leading us from (6) to (7)) so that we apply (12) only in the cases

$$l=1, \dots, \frac{p-1}{2}.$$

So with respect to the component representation (9), we obtain

$$\left\langle \mathfrak{p}(x)^l (1+\mathfrak{v}(x))^{\frac{p-1}{2}-l} \right\rangle_k = 0 \qquad \left(l=1, \dots, \frac{p-1}{2};\ \frac{r+1}{2} \leqq pk+l \leqq r-1\right).$$

According to (1), here all the $k$ that satisfy

$$\frac{s+1}{2} \leqq k \leqq s-1$$

are to be admitted and only these. Thus

$$\left\langle \mathfrak{p}(x)^l (1+\mathfrak{v}(x))^{\frac{p-1}{2}-l} \right\rangle_k = 0 \qquad \left(l=1, \dots, \frac{p-1}{2};\ \frac{s+1}{2} \leqq k \leqq s-1\right). \tag{13}$$

Now we introduce the polynomials

$$\mathfrak{u}(x), \mathfrak{a}(x), \mathfrak{b}_1(x), \dots, \mathfrak{b}_{\frac{p-1}{2}}(x) \tag{14}$$

as follows. We first put

$$\mathfrak{u}(x) = (\mathfrak{p}(x))_{x^s}, \quad \mathfrak{a}(x) = (1+\mathfrak{v}(x))_{x^s}. \tag{15}$$

Then

$$\mathfrak{b}_l(x) = \left(\mathfrak{u}(x)^l \mathfrak{a}(x)^{\frac{p-1}{2}-l}\right)_{x^s} \quad \left(l = 1, \ldots, \frac{p-1}{2}\right). \tag{16}$$

Then ($2_2$) and (3) entail

$$\mathfrak{u}(0)=1, \quad \mathfrak{a}(0)=1. \tag{17}$$

We shall show that

$$\mathfrak{u}^{\circ} \leqq s-1, \quad \mathfrak{a}^{\circ} \leqq \frac{s-1}{p-1}, \quad \mathfrak{b}_l^{\circ} \leqq \frac{s-1}{2} \quad \left(l = 1, \ldots, \frac{p-1}{2}\right) \tag{18}$$

and

$$\mathfrak{u}(x)^l \mathfrak{a}(x)^{\frac{p-1}{k}-l} \equiv \mathfrak{b}_l(x) \pmod{x^s} \quad \left(l = 1, \ldots, \frac{p-1}{2}\right). \tag{19}$$

Of these equations ($18_1$) is trivially valid, as a consequence of ($15_1$). From (10) and ($15_2$) it follows that the inequality ($18_2$) is also true. Since (13) retains its validity after the substitutions

$$\mathfrak{p}(x) \to \mathfrak{u}(x), \quad 1+\mathfrak{v}(x) \to \mathfrak{a}(x),$$

it follows from this that ($18_3$) holds. Finally, (19) follows from (16).

Now by ($1_1$), (14), (17), (18) and (19) the premises of Proposition 13 are satisfied; we therefore have

$$\mathfrak{u}^{\circ} \leqq \frac{s-1}{p-1}.$$

From (15), this inequality can be replaced by

$$(\mathfrak{p}(x))^{\circ}_{x^s} = \frac{s-1}{p-1}.$$

This proves **b** and, consequently, also Proposition 14.

## § 18. The solution of the *a*, *b*-polynomial equation

Thanks to Proposition 14 we are now at last in a position to determine all the solutions $a(x)$, $b(x)$ of the $a$, $b$-polynomial equation

$$xa(x)^p + b(x)^p = x^r a(x)^{p-2} + a(x)^{p-3} b(x) \tag{1}$$

$$\left(r = \frac{q}{p};\ a(x) \in \underset{1}{\mathsf{F}}[x],\ b(x) \in \mathsf{F}[x];\ b^{\circ} \leqq a^{\circ} = \frac{r-1}{2}\right).$$

As our first (and decisive) step in this direction we prove the very strong assertion that $a(x)$ must be the $\frac{p-1}{2}$-th power of a polynomial. (As $a(x)$ is monic, it follows that it is the $\frac{p-1}{2}$-th power of a unique monic polynomial from $\mathsf{F}[x]$.) Once we have obtained this result, the above problem will give us no further difficulty.

The proof of our assertion will also not be difficult (owing to Proposition 14), but we have not found a direct way of doing it. (Cf. Note 3 at the end of this section.) But first we shall need the following simple Proposition.

PROPOSITION 15. *If the pair* $a(x)$, $b(x)$ *is a solution of the a, b-polynomial equation* (1), *then the same holds for all pairs*

$$a(x+\varrho),\quad \varrho^r(x+\varrho)+b(x+\varrho) \qquad (\varrho \in \mathsf{F}). \tag{2}$$

To prove this, we carry out, in (1), the translation $x \to x+\varrho$:

$$(x+\varrho)a(x+\varrho)^p+b(x+\varrho)^p = (x^r+\varrho^r)a(x+\varrho)^{p-2}+a(x+\varrho)^{p-3}b(x+\varrho).$$

As

$$\varrho(=\varrho^q)=(\varrho^r)^p,$$

this equation can be written in the form

$$xa(x+\varrho)^p+\big(\varrho^r a(x+\varrho)b(x+\varrho)\big)^p =$$
$$= x^r a(x+\varrho)^{p-2}+a(x+\varrho)^{p-3}\big(\varrho^r a(x+\varrho)+b(x+\varrho)\big)$$

whence Proposition 15 follows.

We shall say rather unprecisely, that (with fixed $a(x)$, $b(x)$) all the solutions (2) form a *translation class* (of the set) of all solutions of the *a*, *b*-polynomial equation,

COROLLARY. *In each translation class* (2) *of the solutions of the a, b-polynomial equation there is a unique representative consisting of monic polynomials of equal degree.*

For, if $\beta$ denotes the coefficient of $x^{\frac{r-1}{2}}$ in $b(x)$, then in the polynomial $(2_2)$ the coefficient of the same power is equal to

$$\varrho^r+\beta.$$

Since $\varrho \to \varrho^r$ is an automorphism, and therefore $\varrho \to \varrho^r+\beta$ is a permutation of $\mathsf{F}$, the corollary follows from this and from Proposition 15.

Accordingly a solution $a(x)$, $b(x)$ of the *a*, *b*-polynomial equation is said to be monic if $b^\circ = a^\circ$ and, besides $a(x)$, also $b(x)$ is monic.

Now we intend to prove the above assertion. To this end let us consider a solution $a(x)$, $b(x)$ of (1). We have to prove that $a(x)$ is the $\frac{p-1}{2}$-th power of a polynomial. By virtue of Proposition 15 and of the corollary, we may confine ourselves to a normed solution and we therefore write

$$b^{\circ}=a^{\circ}, \quad b(x)\underset{1}{\in}\mathsf{F}[x]. \tag{3}$$

From $a(x)$, $b(x)$ we pass over to the reciprocal polynomials

$$x^{\frac{r-1}{2}}a\left(\frac{1}{x}\right), \quad x^{\frac{r-1}{2}}b\left(\frac{1}{x}\right),$$

retaining, however, for these the notation $a(x)$, $b(x)$. Because of the divisibility

$$\frac{p-1}{2}\,\Big|\,\frac{r-1}{2},$$

it is clear that it suffices to prove that the "new" $a(x)$ is a $\frac{p-1}{2}$-th power in $\mathsf{F}[x]$.

According to § 14, the reciprocal $a, b$-polynomial equation

$$a(x)^p+xb(x)^p = a(x)^{p-2}+x^r a(x)^{p-3}b(x) \tag{4}$$

$$\left(r=\frac{q}{p};\ a(x), b(x)\in\mathsf{F}[x];\ a^{\circ}, b^{\circ}\leqq\frac{r-1}{2};\ a(0)=b(0)=1\right)$$

holds for the new $a(x)$, $b(x)$. We have here used the fact that (3) holds for the "old" $a(x)$, $b(x)$.

We consider (4) as a special case of

$$a(x)^p+xb(x)^p = a(x)^{p-2}+x^r a(x)^{p-3}b(x) \tag{5}$$

$$\left(r=\frac{q}{p};\ a(x), b(x)\in\mathsf{F}[[x]];\ a(0)=b(0)=1\right).$$

The reader will see that (5) is a special case of the $a, b$-power series equation, differing from the general case only in respect of the additional condition $b(0)=1$. Therefore, by Theorem 7

$$\begin{gathered} a(x)=\mathfrak{q}(x)^{\frac{p-1}{2}}, \quad b(x)=\mathfrak{p}(x)\mathfrak{q}(x)^{\frac{p-3}{2}}, \\ \mathfrak{q}(x)=1+\sum_{0\leqq i<\infty}(x\mathfrak{p}(x)^p-x^r\mathfrak{p}(x)^{p^i}, \end{gathered} \tag{6}$$

where the uniformizing parameter $\mathfrak{p}(x)$ satisfies the conditions

$$\mathfrak{p}(x) \in \mathsf{F}[[x]], \quad \mathfrak{p}(0) = 1; \tag{7}$$

for, since $(6_3)$ implies that $\mathfrak{q}(0)=1$, the validity of $(7_2)$ follows from the above additional condition and from $(6_2)$.

For the case $q=p$ we have ($r=1$ and thus) $a(x)=1$, so that the assertion now holds trivially. Henceforth let us assume that

$$q \geqq p^2 \tag{8}$$

We shall write:

$$r = \frac{q}{p}, \quad s = \frac{q}{r^2} \tag{9}$$

(the first of which was introduced in (1)).

We put

$$\mathfrak{v}(x) = \sum_{0 \leqq i < \infty} (x\mathfrak{p}(x)^p)^{p^i} \tag{10}$$

(We must emphasize that the proof to be carried out will be based on (4) and (6)—(10), whereas (5) will henceforth be disregarded.)

By (4) $a^\circ \leqq \dfrac{r-1}{2}$ holds, and hence $\langle a(x) \rangle_k = 0$ for $k \geqq \dfrac{r+1}{2}$, which we shall use, however, only for $k = \dfrac{r+1}{2}, \ldots, r-1$. From $(6_1)$ we are thus led to the equation

$$\left\langle \mathfrak{q}(x)^{\frac{p-1}{2}} \right\rangle_{\frac{r+1}{2}, r-1} = 0. \tag{11}$$

Here $\mathfrak{q}(x)$ will only be considered mod $x^r$, and hence we obtain, by $(6_3)$ and (10),

$$\left\langle (1 + \mathfrak{v}(x))^{\frac{p-1}{2}} \right\rangle_{\frac{r+1}{2}, r-1} = 0. \tag{12}$$

According to (7), (9), (10) and (12), the premises of Proposition 14 are satisfied, whence it follows that the residue

$$(\mathfrak{p}(x))_{x^s} \tag{13}$$

is a $p$-polynomial. The same thing holds for its $p$-th power

$$(\mathfrak{p}(x)^p)_{x^r}$$

and therefore for all powers

$$(x\mathfrak{p}(x)^p)_{x^r}^{p^i}$$

and *a fortiori* for the residues

$$\left((x\mathfrak{p}(x)^p)^{p^i}\right)_{x^r} \qquad (i=0, 1, \ldots).$$

Hence, and from (10), we infer that $\left((\mathfrak{v}(x)\right)_{x^r}$, and accordingly also

$$\left(1+\mathfrak{v}(x)\right)_{x^r}$$

is a $p$-polynomial. For its degree the inequality

$$\left(1+\mathfrak{v}(x)\right)_{x^r}^{\circ} \leqq \frac{r-1}{p-1} \tag{14}$$

must hold.

On the other hand, it follows from $(6_1)$, $(6_3)$ and (10) first of all that

$$a(x) \equiv \left(1+\mathfrak{v}(x)\right)^{\frac{p-1}{2}} \quad (\text{mod } x^r)$$

and then from this congruence that

$$a(x) \equiv \left(1+\mathfrak{v}(x)\right)_{x^r}^{\frac{p-1}{2}} \quad (\text{mod } x^r).$$

The left-hand side is of degree $\dfrac{r-1}{2}$ (since, according to our assumption, $a(x)$, $b(x)$ is a solution of (4)). By (14), the right-hand side is of at most the same degree, whence it follows that the two sides are equal. This completes the proof of the assertion,

We point out that the proof carried out here establishes merely the existence of a polynomial whose $\dfrac{p-1}{2}$-th power is equal to $a(x)$. We do not see any possibility of extending the proof in such a way that it might yield more. However, using the existence result that we already have, we shall be able to get complete results applying elementary techniques.

For this purpose we again consider a solution $a(x)$, $b(x)$ of (1). By what we have already proved, we can write

$$a(x) = c_1(x)^{\frac{p-1}{2}} \qquad \left(c_1(x) \underset{1}{\in} \mathsf{F}[x],\ c_1^{\circ} = \frac{r-1}{p-1}\right). \tag{15}$$

(The notation $c_1(x)$ is most appropriate, as we shall see later.) Substituting into (1)

$$xc_1(x)^{p\frac{p-1}{2}} + b(x)^p = x^r c_1(x)^{(p-2)\frac{p-1}{2}} + c_1(x)^{(p-3)\frac{p-1}{2}} b(x). \tag{16}$$

We shall show that

$$c_1(x)^{\frac{p-3}{2}} \Big| b(x). \tag{17}$$

In the proof, $b(x)\neq 0$ may be assumed. Let $u(x)$ be a linear factor of $c_1(x)$ from $\bar{\mathsf{F}}[x]$. We define $k$ and $l$ by

$$u(x)^k \| c_1(x), \quad u(x)^l \| b(x).$$

It suffices to prove that

$$l \geqq \frac{p-3}{2}k.$$

Let us assume this assertion to be false. Then there is a $u(x)$ for which

$$l < \frac{p-3}{2}k$$

holds. It follows that the multiplicity with which $u(x)$ is contained as a factor in the left- and right-hand sides of (16) are the following:

$$pl, (p-3)\frac{p-1}{2}k+l.$$

These must be equal, so that

$$l = \frac{p-3}{2}k.$$

This contradiction proves (17).

Therefore we can put

$$b(x) = c_0(x)c_1(x)^{\frac{p-3}{2}} \qquad \left(c_0(x)\in\mathsf{F}[x],\ c_0^\circ \leqq \frac{r-1}{p-1}\right), \tag{18}$$

with a suitable $c_0(x)$.

After substitution of (18) into (16) and division by a power of $c_1(x)$ we obtain the simple equation

$$xc_1(x)^p + c_0(x)^p = x^r c_1(x) + c_0(x). \tag{19}$$

With this partial result we are obviously very near to solving (1). (19) could be solved by coefficient comparison but this would be rather tedious. A simple idea, however, will lead us straight to our goal; this idea consists in using instead of the pair of polynomials $c_0(x)$, $c_1(x)$, the polynomial

$$c(x) = x^r c_1(x) + c_0(x), \tag{20}$$

composed of $c_0(x)$ and $c_1(x)$. (20) will occasionally be called the *kernel* of the *a*, *b*-polynomial equation; this nomenclature was a reasonable choice in view of the fact that, conversely, $c(x)$ clearly determines $c_0(x)$ and $c_1(x)$ uniquely by means of the formulae

$$c_0(x) = (c(x))_{x^r}, \quad c_1(x) = \left[\frac{c(x)}{x^r}\right]. \tag{21}$$

These $c_0(x)$ and $c_1(x)$ are called, for brevity (in view of (20)), the two (*one-step*) *constituents* of the kernel $c(x)$. The point is then that we have stressed the kernel, $c(x)$, instead of its constituents $c_0(x)$, $c_1(x)$. (It will soon transpire that $c_0(x)$ and $c_1(x)$ are actually the non-zero reduced components of $c(x)$.)

Now (20) implies that

$$c(x)^p = x^q c_1(x)^p + c_0(x)^r,$$

and consequently by (19) we have the equation

$$c(x)^p - c(x) = (x^q - x)c_1(x)^r \tag{22}$$

and thus the divisibility

$$x^q - x | c(x)^p - c(x). \tag{23}$$

By Theorem 1, this means that $c(x)$ is an $(\mathsf{F}, \mathsf{F}_p)$-polynomial, i.e. a rational-valued polynomial.

Moreover, it should be noted that, by (15), (18) and (20), the conditions

$$c(x) \underset{1}{\in} \mathsf{F}[x], \quad c^\circ = \frac{q-1}{p-1} \tag{24}$$

are satisfied. (Here $r + \dfrac{r-1}{p-1} = \dfrac{q-1}{p-1}$ has been used.)

Thanks to (19), however, (20) can be written in the ("second") form

$$c(x) = xc_1(x)^p + c_0(x)^p. \tag{25}$$

This implies that the 0-th $p$-adic numeral of the elements of the exponent set $\mathfrak{E}(c)$ can only be either 0 or 1. Since, on the other hand, $(24_2)$ implies that $c^\circ < q$, it follows from Theorem 3 because of (23) that $\mathfrak{E}(c)$ is the union of $p$-cycles, and thus the preceding statement must hold for all $p$-adic numerals of the elements of $\mathfrak{E}(c)$. This means that $c(x)$ is $p$-linear, i.e. (since $c^\circ \geqq 1$)

$$c^{\circ p} = 1. \tag{26}$$

(By virtue of (25), it even follows that for $i=0, 1$ just $c_i(x)$ is the $i$-th reduced component of $c(x)$ and the other reduced components of $c(x)$ vanish; we have noticed this only by the way, however, since we want to use the notation introduced at (21).

It is to see that, by (23), (24) and (26), $c(x)$ is a stem polynomial. (It must repeatedly be emphasized that the stem polynomials are completely known, thanks to § 4.)

Conversely, let us suppose that $c(x)$ is a stem polynomial, i.e. that (23), (24) and (26) are satisfied. We shall prove that the formulae (15), (18) and (21) then supply a solution of the $a$, $b$-polynomial equation (1).

By (21), we have (20). Hence and from (23) we deduce, using $pr=q$, the congruence

$$xc_1(x)^p+c_0(x)^p \equiv x^r c_1(x)+c_0(x) \pmod{x^q-x}. \tag{27}$$

From (21), (24) and (26) the inequality

$$c_0^\circ,\, c_1^\circ \leqq \frac{r-1}{p-1} \tag{28}$$

follows. Thus both sides of (27) are of degree

$$1+p\frac{r-1}{p-1}\left(=r+\frac{r-1}{p-1}=\frac{q-1}{p-1}<q\right),$$

and therefore equation (19) follows from (27). Multiplying the latter by

$$c_1(x)^{p\frac{p-3}{2}}$$

equation (1) follows, by (15) and (18). On account of (15), (18), (24) and (28) it is also clear that the accessory conditions indicated in (1) are satisfied. We have thus proved

THEOREM 8. *All solutions of the a, b-polynomial equation*

$$xa(x)^p+b(x)^p = x^r a(x)^{p-2}+a(x)^{p-3}b(x) \tag{29}$$

$$\left(r=\frac{q}{p};\ a(x)\underset{1}{\in}\mathsf{F}[x],\ b(x)\in\mathsf{F}[x];\ b^\circ \leqq a^\circ = \frac{r-1}{2}\right)$$

*are given by the solution formulae*

$$a(x)=c_1(x)^{\frac{p-1}{2}},\quad b(x)=c_0(x)c_1(x)^{\frac{p-3}{2}},\quad c_0(x)=(c(x))_{x^r},\quad c_1(x)=\left[\frac{c(x)}{x^r}\right], \tag{30}$$

*where $c(x)$ runs through all stem polynomials for p, i.e. all polynomials with*

$$c(x)\underset{1}{\in}\mathsf{F}[x],\quad c^\circ=\frac{q-1}{p-1},\quad c^{\circ p}=1,\quad x^q-x\,|\,c(x)^p-c(x). \tag{31}$$

NOTE 1. Since these $c(x)$ are explicitly determined by the special case $P=p$ of Theorem 3 (cf. § 4), the solutions of the $a$, $b$-polynomial equation are known thanks to Theorem 8. We see that the kernel $c(x)$ of this equation here plays the role of a "uniformizing parameter" (and coincides with the concept of the stem polynomial for $p$).

NOTE 2. Theorem 6 and Proposition 12 can be reworded, of course, by substituting $a(x)$, $b(x)$ from Theorem 8. More complete results can be obtained in this way. As it is only in an auxiliary role that we are concerned with the $\lambda$-differential equation, we shall not carry out the (rather easy) rewording of Theorem 6, but that of Proposition 12 will form the subject of the next section,

NOTE 3. Before finding the above difficult proof (based on §§ 15, 16 and 17) of Theorem 8 we made many attempts to use elementary approaches of which we give a brief account. By differentiation (1) is easily transformed into

$$(a(x)^3+2x^r a'(x))^\circ \leqq a^\circ.$$

An equally simple consideration of divisibility gives us that

$$a(x)=u(x)^p v(x)^{\frac{p-1}{2}}$$

or

$$a(x)=x^{1+3k}u(x)^p v(x)^{\frac{p-1}{3}} \qquad (k=0, 1, \ldots)$$

must hold in (1), where $u(x)$, $v(x)$ are monic polynomials from $\mathsf{F}[x]$, prime to one another and, in the second case, also to $x$. Although both conditions would seem to be very incisive, we have not been able to shorten the proof of Theorem 8 by applying them.

NOTE 4. For the purpose of a later application we must stress the part of the above result that for the constituents $c_0(x)$, $c_1(x)$, as defined by (21), of a stem polynomial $c(x)$ for $p$, the relations (20), (25) and also (22), all hold.

## § 19. Reduction of the marginal case of Problem II to that of Problem III

We take Proposition 12 as our starting-point. This states that the solutions of Problem II, for the marginal case, are given by

$$f(x) = \left(x+\frac{b(x^p)}{a(x^p)}\right)\left(a(x^p)\left(x+\frac{b(x^p)}{a(x^p)}\right)^{\frac{p-1}{2}}+\sigma\right)\left(a(x^p)\left(x+\frac{b(x^p)}{a(x^p)}\right)^{\frac{p-1}{2}}+\sigma\tau\right) \qquad (1)$$

$$(\sigma=\pm 1;\ \tau=0, 1),$$

where the pair $a(x)$, $b(x)$ has to run through all the solutions of the $a$, $b$-polynomial equation for which this polynomial quotient $f(x)$ is integral and fully reducible. On the other hand, according to Theorem 8, all the $a(x)$, $b(x)$ are given by the formulae

$$a(x) = c_1(x)^{\frac{p-1}{2}}, \quad b(x) = c_0(x)c_1(x)^{\frac{p-3}{2}}, \quad c_0(x) = (c(x))_{x^r}, \quad c(x) = \left[\frac{c(x)}{x^r}\right] \tag{2}$$

$$\left(r = \frac{q}{p}\right),$$

where (the kernel) $c(x)$ is characterized by the conditions

$$c(x) \underset{1}{\in} \mathsf{F}[x], \quad c^{\circ} = \frac{q-1}{p-1}, \quad c^{\circ_p} = 1, \quad x^q - x \mid c(x)^p - c(x), \tag{3}$$

i.e. it means all the stem polynomials for $p$. (The possibility of the explicit determination of these $c(x)$, given by Theorem 3, will here not yet be relevant to our purpose.)

We substitute (2) into (1), thus making considerable progress for the marginal case of Problem II.

We begin by mentioning that the relations

$$c(x) = x^r c_1(x) + c_0(x) = xc_1(x)^p + c_0(x)^p \tag{4}$$

and

$$c(x)^p - c(x) = (x^q - x)c_1(x)^p \tag{5}$$

follow from (3) (see § 18, Note 4).

Now we first of all deduce from $(2_1)$, $(2_2)$ that

$$x + \frac{b(x^p)}{a(x^p)} = x + \frac{c_0(x^p)}{c_1(x^p)} = \frac{xc_1(x^p) + c_0(x^p)}{c_1(x^p)}. \tag{6}$$

Hence and again from $(2_1)$ the relation

$$a(x^p)\left(x + \frac{b(x^p)}{a(x^p)}\right)^{\frac{p-1}{2}} = (xc_1(x^p) + c_0(x^p))^{\frac{p-1}{2}} \tag{7}$$

also follows.

In order that the substitution of (6) and (7) into (1) should lead to a simple result, we apply the automorphism $\xi \rightarrow \xi^p$ of $\mathsf{F}$ to the coefficients of $c(x)$. This is allowed, as thereby all the stem polynomials $c(x)$ undergo only permutations among themselves. So $c_0(x^p)$, $c_1(x^p)$ become $c_0(x)^p$, $c_1(x)^p$, and the right-hand side of (6) and (7) change, by (4), to

$$\frac{c(x)}{c_1(x)^p} \quad \text{and} \quad c(x)^{\frac{p-1}{2}},$$

respectively. Thus from (1) we obtain

$$f(x) = \frac{c(x)}{c_1(x)^p}\left(c(x)^{\frac{p-1}{2}} + \sigma\right)\left(c(x)^{\frac{p-1}{2}} + \sigma\tau\right) \qquad (\sigma = \pm 1;\ \tau = 0, 1). \tag{8}$$

A further considerable simplification arises with the help of equation (5) which will be written in the form

$$\frac{c(x)}{c_1(x)^p} = \frac{x^q - x}{c(x)^{p-1} - 1}.$$

Substituting this into (8) and replacing $\sigma$ by $-\sigma$ we obtain after a slight reduction

$$f(x) = \frac{x^q - x}{c(x)^{\frac{p-1}{2}} + \sigma}\left(c(x)^{\frac{p-1}{2}} - \sigma\tau\right) \qquad (\sigma = \pm 1;\ \tau = 0, 1). \tag{9}$$

As the $c(x)$ are completely known, the only questions remaining to be discussed are to decide whether $f(x)$ is integral and fully reducible. Both questions can be simplified as follows.

In (9) the denominator is prime to the second factor of the right-hand side. Therefore $f(x)$ is integer if and only if

$$c(x)^{\frac{p-1}{2}} + \sigma \,\big|\, x^q - x \tag{10}$$

holds. This, however, means, together with (3), that $c(x)$ is a solution of Problem III, for the pair

$$(P =)p, \quad (\varphi(x) =)x^{\frac{p-1}{2}} + \sigma \qquad (\sigma = \pm 1).$$

Moreover, if this is true, then the first factor on the right-hand side of (9) is (as a divisor of $x^q - x$) also fully reducible. Thus, for $f(x)$ itself to be fully reducible it is necessary and sufficient that the same should hold for the second factor. We have thus proved the

MAIN LEMMA. *The only solutions of the marginal case of Problem II are the*

$$f(x) = \frac{x^q - x}{c(x)^{\frac{p-1}{2}} + \sigma}\left(c(x)^{\frac{p-1}{2}} - \sigma\tau\right) \qquad (\sigma = \pm 1;\ \tau = 0, 1) \tag{11}$$

*in which* $c(x)$ *is a solution of Problem III, belonging to the pair*

$$(P=)p, \quad (\varphi(x)=)x^{\frac{p-1}{2}}+\sigma, \tag{12}$$

*and the second factor on the right-hand side of* (11) *is fully reducible.*

NOTE 1. Of the two problems concerning $c(x)$ that, according to the Main Lemma, we have to discuss, the second appears to be much the more difficult since it refers "only" to the full reducibility, whereas in the first much more is required, namely also the absence of multiple factors. It would therefore be only natural to commence with the first problem, i.e. with the investigation of Problem III. This program will be carried out in Chapter IV (not only for the pair (12) but in complete generality, for the above reasons). This too will be very difficult but in Chapter V it will turn out—as a pleasant surprise—that the second problem (still more difficult in itself) will herewith automatically be solved. (Exceptions are the cases $p=3, 5, 7$.) This means that to solve the marginal case of Problem II we shall have to deal in the main only with Problem III.

NOTE 2. It will be shown that it is sufficient to treat the Main Lemma for $\sigma=1$, by which it should be understood that all solutions of the marginal case of Problem II are given by (11) even with this restriction if no notice is taken of rotation-equivalent solutions; moreover, it will be shown that the same thing holds for $\sigma=-1$. It should be noted that for all $\xi\in\mathsf{F}$ all the $c(x)$ undergo a permutation among themselves as a result of the transformation

$$c(x)\to\bar{c}(x)=\xi^{-c^{\circ}}c(\xi x). \tag{13}$$

In other words, this means the rotation invariance of the stem polynomials (where we have used the fact that they are monic). This invariance is obvious for the properties $(3_1)$, $(3_2)$, $(3_3)$ and it also holds for the property $(3_4)$, as by Theorem 1 this last is equivalent to the property of $c(x)$ that it is rational-valued and this property extends to $\bar{c}(x)$ by virtue of $\xi^{c^{\circ}}\in\mathsf{F}_p$ (cf. $(3_2)$). (This invariance could, of course, be verified directly.) It follows that the same $f(x)$ are given by (11) and by

$$f(x)=\frac{x^q-x}{\bar{c}(x)^{\frac{p-1}{2}}+\sigma}\left(\bar{c}(x)^{\frac{p-1}{2}}-\sigma\tau\right) \tag{14}$$

Now, in view of the invariance of Problem II and of $\xi^q=\xi$ we can here go over to the rotation equivalent solutions

$$\xi f\left(\frac{x}{\xi}\right).$$

Retaining the above notation, (14) then turns into

$$f(x) = \frac{x^q - x}{c(x)^{\frac{p-1}{2}} + \xi^{\frac{q-1}{2}} \sigma} \left( c(x)^{\frac{p-1}{2}} - \xi^{\frac{q-1}{2}} \sigma\tau \right). \tag{15}$$

Finally, as the arbitrary $\xi$ can be chosen so that the power

$$\xi^{\frac{q-1}{2}}$$

should be equal to $\sigma$ and also so that it should be equal to $-\sigma$, this proves the assertion. Although in principle a simplification of the Main Lemma is thus achieved, for further purposes this lemma is more suitable in its original form.

NOTE 3. As in (11) the denominator and the second factor of the right-hand side are relatively prime to each other, it is an obvious consequence of the Main Lemma that every solution of the marginal case of Problem II must have exactly $\frac{q+1}{2}$ distinct zeros.

CHAPTER IV

# PROBLEM III

On the one hand, it is possible to give explicitly a whole host, very extended in general, of certain "regular" solutions of Problem III. It turns out, on the other hand, that in many cases no further solutions exist. The results discussed below will suffice to solve the marginal case of Problem II completely (except the cases $q=3^n$, $5^n$, $7^n$ with $n \geqq 4$). Throughout this chapter we shall assume $p \neq 2$, since Problem III (and the marginal case of Problem II) deals only with these values of $p$.

## § 20. A lemma on the greatest common divisor of polynomials with common gap

For our purpose we shall use on several occasions:

LEMMA I. *For the greatest common divisor of polynomials of the form*

$$f(x) = x^k f_1(x) + f_2(x), \quad g(x) = x^k g_1(x) + g_2(x) \tag{1}$$

$$(k \in \mathcal{N};\ f_1(x), f_2(x), g_1(x), g_2(x) \in \mathsf{F}[x])$$

*the divisibility*

$$(f(x), g(x)) \left| \begin{vmatrix} f_1(x) & f_2(x) \\ g_1(x) & g_2(x) \end{vmatrix} \right. \tag{2}$$

*holds.*

COROLLARY. *If in* (1)

$$f_1^\circ, g_1^\circ \leqq i; \quad f_2^\circ, g_2^\circ \leqq j \tag{3}$$

*hold and if the right-hand side of* (2) *does not vanish, then*

$$(f(x), g(x))^\circ \leqq i+j. \tag{4}$$

From (1) the right-hand side of (2) is equal to

$$\begin{vmatrix} f_1(x) & f(x) \\ g_1(x) & g(x) \end{vmatrix},$$

whence Lemma I follows. The corollary is trivial.

NOTE 1. We shall apply Lemma I and its corollary only for

$$k \geqq j+2. \tag{5}$$

Then we have to deal with two polynomials $f(x)$, $g(x)$ in which the coefficients of the powers

$$x^{j+1}, \ldots, x^{k-1}$$

vanish. We may therefore say that these polynomials have a *common gap* of length at least equal to

$$k-j. \tag{6}$$

Since the estimation

$$f^{\circ}, g^{\circ} \leqq i+k \tag{7}$$

now holds (in general, clearly the best one possible) and the difference between the right-hand side of (7) and that of (4) is also $k-j$, indicated in (6), it is clear that in (4) the assumed lacunarity is fully exploited. (There will be no need to treat this more precisely.) The corollary can therefore reasonably be called "strong". Considerable progress will be made each time we use the corollary, without which similar results hardly could be obtained.

NOTE 2. In problems involving a greatest common divisor

$$d(x) = (f(x), g(x)) \quad (f(x), g(x) \in \mathsf{F}[x] \setminus 0)$$

we often adopt the classical device of the Euclidean Algorithm. Suppose that it is used on $f(x)$ and $g(x)$ and let $r_1(x), \ldots, r_m(x)$ denote the successive non-zero residues. We thus arrive at the degree estimates

$$d^{\circ} \leqq r_t^{\circ} \qquad (t=1, \ldots, m)$$

which get stronger and stronger as the index $t$ increases. It follows, nevertheless, from the nature of the Euclidean Algorithm that a proof of the corollary cannot be expected from this approach because it is clear that the algorithm is almost wholly insensitive to the lacunarity of the polynomials chosen at the outset. We direct the reader's attention to the fact that the effective proof of Lemma I given above (and its corollary) is based substantially on the concept of the ideal. (We might put it thus: "Dedekind contra Euclid".)

NOTE 3. One disadvantage of the corollary is that the case when the right-hand side of (2) vanishes is an exception. This exceptional case must be dealt with separately every time we use the corollary.

Of course, Lemma I and its corollary admit of many generalizations, but the form given above will suffice for our purposes.

## § 21. The theorem for four *P*-linear polynomials

As a further aid we shall now make some statements about *P*-linear polynomials.

Following L. Kalmár (see Note 2, below) we shall shortly say that in a subset $T$ of a commutative semigroup $S$ the *theorem for four elements* holds if from the supposition

$$\alpha\delta=\beta\gamma \qquad (\alpha, \beta, \gamma\ \delta \in T) \tag{1}$$

the existence of four further elements $\varkappa, \lambda, \mu, \nu \in T$ such that

$$\alpha=\varkappa\lambda, \quad \delta=\mu\nu, \quad \beta=\varkappa\mu, \quad \gamma=\lambda\nu \tag{2}$$

follows. If the same is true for $T=S$, we shall say that the theorem for four elements holds in $S$. We illustrate the circumstances of the theorem for four elements by Figure 1 in which each of the 4 given elements

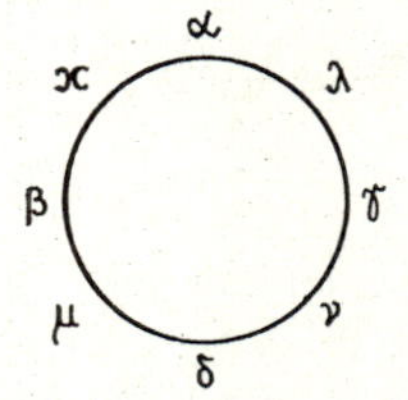


*Fig. 1*

is equal to the product of the two (cyclically) neighbouring elements.

We shall be interested only in the case where a commutative ring replaces $S$. Then the property that is responsible for the validity of the theorem for four elements can be characterized very simply as follows: In a subset $T$ of a commutative ring the validity of the theorem for four elements means that every vanishing determinant

$$\begin{vmatrix} \alpha & \beta \\ \gamma & \delta \end{vmatrix} \qquad (\alpha, \beta, \gamma, \delta \in T)$$

is necessarily of the form

$$\begin{vmatrix} \varkappa\lambda & \varkappa\mu \\ \nu\lambda & \nu\mu \end{vmatrix} \qquad (\varkappa, \lambda, \mu, \nu \in T)$$

(where — as is easily seen — every row and column vector contains a scalar factor).

NOTE 1. In terms of matrices this means the possibility of a product decomposition

$$\begin{pmatrix} \alpha & \beta \\ \gamma & \delta \end{pmatrix} = \begin{pmatrix} \varkappa & 0 \\ \nu & 0 \end{pmatrix} \begin{pmatrix} \lambda & \mu \\ 0 & 0 \end{pmatrix}$$

whenever the left-hand side is singular.

NOTE 2. The theorem for four elements holds, for instance, in all prime factorization rings, in particular in $\mathsf{F}[x]$. A fundamental significance of the theorem for four elements in $\mathscr{Z}$ was discovered by L. Kalmár [4] who called it in this case the "theorem for four numbers" (in German "Vierzahlensatz"). He recognized that it is very easy to find a proof for this theorem and thus produced (*inter alia*) a very simple proof of the fundamental theorem of elementary number theory.

If (1) and (2) are applied to polynomials then we speak "theorem for four polynomials" and if besides in (2) we are allowed to require direct products in place of ordinary products, then we shall talk about the *strong theorem for four polynomials*, instead of the "theorem of four elements".

We shall first prove two preliminary propositions.

PROPOSITION 16. *For P-linear polynomials $u(x), v(x), w(x) \in \mathsf{F}[x]$ with*

$$u(x) = v(x)w(x)$$

*it is always true that*

$$u(x) = v(x) \times w(x)$$

(*consequently at least one factor must lie in* $\mathsf{F}[x^P]$).

In the case $u^\circ \leqq 0$ the assertion holds trivially. In the remaining case $u^\circ \geqq 1$ we suppose the assertion to be true for the smaller $u^\circ$. We shall use the component representations

$$u(x) = u_0(x^P) + xu_1(x^P), \quad v(x) = v_0(x^P) + xv_1(x^P), \quad w(x) = w_0(x^P) + xw_1(x^P)$$

with polynomials $u_0(x), \ldots, w_1(x) \in \mathsf{F}[x]$. It is clear that these polynomials must likewise be $P$-linear and

$$v_1(x)w_1(x) = 0$$

necessarily holds; we may therefore assume, e.g., $v_1(x) = 0$. Then we have simply

$$u_0(x) = v_0(x)w_0(x), \quad u_1(x) = v_0(x)w_1(x).$$

Moreover, we have $u_0^\circ, u_1^\circ < u^\circ$ and therefore the validity of

$$u_i(x) = v_0(x) \times w_i(x) \qquad (i = 0, 1)$$

follows from the induction hypothesis. We obtain

$$u(x) = v_0(x^P) \times w_0(x^P) + v_0(x^P) \times xw_1(x^P),$$

i.e.

$$u(x) = v_0(x^P) \times (w_0(x^P) + xw_1(x^P)) = v(x) \times w(x).$$

Proposition 16 is thus proved.

PROPOSITION 17. *For P-linear polynomials $a(x), b(x), c(x), d(x) \in \mathsf{F}[x]$ with the component representations*

$$a(x) = a_0(x^P) + xa_1(x^P), \ldots, d(x) = d_0(x^P) + xd_1(x^P) \tag{3}$$

$$(a_0(x), a_1(x), \ldots, d_1(x) \in \mathsf{F}[x])$$

*the equation*

$$\begin{vmatrix} a(x) & b(x) \\ c(x) & d(x) \end{vmatrix} = 0 \tag{4}$$

*holds if and only if all the four equations*

$$\begin{vmatrix} a_i(x) & b_j(x) \\ c_i(x) & d_j(x) \end{vmatrix} = 0 \qquad (i, j = 0, 1) \tag{5}$$

*hold or if this is true after the transposition $b(x) \leftrightarrow c(x)$.*

(The necessity of (5) is trivial for $i=j$, as will become manifest also in the course of the proof.)

NOTE 3. This proposition asserts that after expansion, substitution of (3) and carrying out the multiplication the left-hand side of (4) becomes the sum of 8 terms, which can be grouped into four pairs, each having a vanishing sum.

The "if" part of Proposition 17 is trivial. In order to prove the "only if", part, we substitute (3) into (4). After a component comparison we get the three equations

$$\begin{vmatrix} a_0(x) & b_0(x) \\ c_0(x) & d_0(x) \end{vmatrix} = 0, \qquad \begin{vmatrix} a_0(x) & b_1(x) \\ c_0(x) & d_1(x) \end{vmatrix} + \begin{vmatrix} a_1(x) & b_0(x) \\ c_1(x) & d_0(x) \end{vmatrix} = 0,$$

$$\begin{vmatrix} a_1(x) & b_1(x) \\ c_1(x) & d_1(x) \end{vmatrix} = 0.$$

The first and the third constitute just the case $i=j$ $(=0, 1)$ of the conditions (5) and it follows from them that in the second equation, after its rearrangement into

$$a_0(x)d_1(x) + a_1(x)d_0(x) = b_1(x)c_0(x) + b_0(x)c_1(x)$$

the product of both terms of the left-hand side will be equal to that of the terms of the right-hand side. Hence the validity of

$$a_0(x)d_1(x) - a_1(x)d_0(x) = \pm(b_1(x)c_0(x) - b_0(x)c_1(x))$$

follows trivially. Performing the transposition $b(x) \leftrightarrow c(x)$ if necessary, it can be achieved that the plus sign is valid. By adding this equation to the preceding one or by subtraction from it we obtain

$$2a_0(x)d_1(x) = 2b_1(x)c_0(x), \quad 2a_1(x)d_0(x) = 2b_0(x)c_1(x).$$

As we may divide by 2 (because $p \neq 2$), the result is just the part $i=j$ of the conditions (5). This completes the proof of Proposition 17.

LEMMA II. *For P-linear polynomials* $a(x), b(x), c(x), d(x) \in \mathsf{F}[x]$ *the strong theorem for four polynomials holds, i.e. from*

$$a(x)d(x) = b(x)c(x) \tag{6}$$

*follows the existence of P-linear polynomials* $\bar{a}(x), \bar{b}(x), \bar{c}(x), \bar{d}(x) \in \mathsf{F}[x]$ *such that*

$$a(x) = \bar{a}(x) \times \bar{b}(x), \quad d(x) = \bar{c}(x) \times \bar{d}(x), \quad b(x) = \bar{a}(x) \times \bar{c}(x), \tag{7}$$
$$c(x) = \bar{b}(x) \times \bar{d}(x).$$

To prove this, we remark that, because of the assumed $P$-linearity of the left-hand sides of the equations (7), it suffices, by Proposition 16, to prove the validity of the "ordinary" theorem for four polynomials, i.e. the fact that

$$a(x) = \bar{a}(x)\bar{b}(x), \quad d(x) = \bar{c}(x)\bar{d}(x), \quad b(x) = \bar{a}(x)\bar{c}(x), \quad c(x) = \bar{b}(x)\bar{d}(x) \tag{8}$$

can be satisfied by $P$-linear factors $\bar{a}(x), \ldots, \bar{d}(x) \in \mathsf{F}[x]$.

We shall first dispose of the case in which one of the given polynomials is constant. By symmetry, e.g., $a(x) \in \mathsf{F}$ may be assumed. If we assume that $a(x)=0$, then it follows that $b(c)\, c(x)=0$, whence we may take $b(x)=0$. Then (8) is satisfied by

$$\bar{a}(x)=0, \quad \bar{b}(x)=c(x), \quad \bar{c}(x)=d(x), \quad \bar{d}(x)=1.$$

If $a(x) \in \mathsf{F} \setminus 0$, then

$$\bar{a}(x)=1, \quad \bar{b}(x)=a(x), \quad \bar{c}(x)=b(x), \quad \bar{d}(x)=a(x)^{-1}c(x)$$

will do.

We still have to accomplish the proof for the case $a^\circ, \ldots, d^\circ \geqq 1$. Suppose the assertion to be true for the smaller sums

$$a^\circ + d^\circ (= b^\circ + c^\circ).$$

We use the component representations

$$a(x) = a_0(x^P) + xa_1(x^P), \ \ldots, \ d(x) = d_0(x^P) + xd_1(x^P), \tag{9}$$

where the 8 polynomials $a_0(x), \ldots, d_i(x) \in \mathsf{F}[x]$ are likewise $P$-linear.

We shall now give the proof for the case where one of the four polynomials $a_0(x), \ldots, d_0(x)$ vanishes. From (6) and (9) it follows in the simplest way from Proposition 17 that

$$\begin{vmatrix} a_0(x) & b_0(x) \\ c_0(x) & d_0(x) \end{vmatrix} = 0. \tag{10}$$

Therefore either a row or a column must vanish on the left. By symmetry it may be assumed that the first row vanishes; then, in view of $(9_1)$ and $(9_2)$, the quotients

$$\frac{a(x)}{x}\,(= a_1(x^P)), \quad \frac{b(x)}{x}\,(= b_1(x^P))$$

are *P*-linear polynomials. On the other hand, (6) implies that

$$\frac{a(x)}{x}\,d(x) = \frac{b(x)}{x}\,c(x)$$

and thus the induction hypothesis implies the existence of *P*-linear polynomials $\bar{a}(x), \ldots, \bar{d}(x) \in \mathsf{F}[x]$ such that

$$\frac{a(x)}{x} = \bar{a}(x)\bar{b}(x), \quad d(x) = \bar{c}(x)\bar{d}(x), \quad \frac{b(x)}{x} = \bar{a}(x)\bar{c}(x), \quad c(x) = \bar{b}(x)\bar{d}(x).$$

The $x$-component of $\bar{a}(x)$ is clearly equal to 0 (so the same must hold for $\bar{b}(x)$ and $\bar{c}(x)$). The product $x\bar{a}(x)$ is therefore certainly *P*-linear. Further, writing the first and the third of the above equations in the form

$$a(x) = x\bar{a}(x)\bar{b}(x), \quad b(x) = x\bar{a}(x)\bar{c}(x),$$

we see that (8) is satisfied by $x\bar{a}(x)$ in place of $\bar{a}(x)$. Thus Lemma II is proved for this case.

We may henceforth assume that $a_0(x), \ldots, d_0(x) \neq 0$. Complete use of Proposition 17 enables us to assume (5) without loss of generality. We consider first the case $i=j=0$, i.e. equation (10) from which, by induction, the existence of four *P*-linear polynomials $a_2(x), \ldots, d_2(x) \in \mathsf{F}[x] \setminus 0$ with

$$\begin{aligned} a_0(x) &= a_2(x)b_2(x), & d_0(x) &= c_2(x)d_2(x), \\ b_0(x) &= a_2(x)c_2(x), & c_0(x) &= b_2(x)d_2(x) \end{aligned} \tag{11}$$

can be inferred. Accordingly, the cases $i=0, j=1$ and $i=1, j=0$ of (5), after dividing by $b_2(x)$ or $c_2(x)$, respectively, become the equations

$$\begin{vmatrix} a_2(x) & b_1(x) \\ d_2(x) & d_1(x) \end{vmatrix} = 0, \qquad \begin{vmatrix} a_1(x) & a_2(x) \\ c_1(x) & d_2(x) \end{vmatrix} = 0. \tag{12}$$

(As for the remaining case $i=j=1$ of (5), it will not be used.)

By induction, the existence of four $P$-linear polynomials $a_3(x), \ldots, d_3(x) \in \in \mathsf{F}[x]$ with

$$a_2(x) = a_3(x) b_3(x), \quad d_1(x) = c_3(x) d_3(x),$$
$$b_1(x) = a_3(x) c_3(x), \quad d_2(x) = b_3(x) d_3(x) \tag{13}$$

follows from $(12_1)$. It should be noted that, from $(11_1)$, $a_2(x) \neq 0$, and thus, by $(13_1)$,

$$b_3(x) \neq 0.$$

By substituting from $(13_1)$ and $(13_4)$ into $(12_2)$ and dividing by $b_3(x)$

$$\begin{vmatrix} a_1(x) & a_3(x) \\ c_1(x) & d_3(x) \end{vmatrix} = 0 \,.$$

Consequently, the existence of four $P$-linear polynomials $a_4(x), \ldots, d_4(x) \in \mathsf{F}[x]$ with

$$a_1(x) = a_4(x) b_4(x), \quad d_3(x) = c_4(x) d_4(x),$$
$$a_3(x) = a_4(x) c_4(x), \quad c_1(x) = b_4(x) d_4(x) \tag{14}$$

follows by induction.

Now by consecutive substitution of (11), (13), (14) we obtain from (9) the formulae

$$\begin{aligned} a(x) &= a_0(x^P) + x a_1(x^P) = a_2(x^P) b_2(x^P) + x a_1(x^P) = \\ &= a_3(x^P) b_3(x^P) b_2(x^P) + x a_1(x^P) = a_4(x^P)[c_4(x^P) b_3(x^P) b_2(x^P) + x b_4(x^P)], \\ b(x) &= b_0(x^P) + x b_1(x^P) = a_2(x^P) c_2(x^P) + x b_1(x^P) = \\ &= a_3(x^P)[b_3(x^P) c_2(x^P) + x c_3(x^P)] = a_4(x^P) c_4(x^P)[b_3(x^P) c_2(x^P) + x c_3(x^P)], \\ c(x) &= c_0(x^P) + x c_1(x^P) = b_2(x^P) d_2(x^P) + x c_1(x^P) = \\ &= b_2(x^P) b_3(x^P) d_3(x^P) + x c_1(x^P) = d_4(x^P)[b_2(x^P) b_3(x^P) c_4(x^P) + x b_4(x^P)], \\ d(x) &= d_0(x^P) + x d_1(x^P) = c_2(x^P) d_2(x^P) + x d_1(x^P) = \\ &= d_3(x^P)[c_2(x^P) b_3(x^P) + x c_3(x^P)] = c_4(x^P) d_4(x^P)[c_2(x^P) b_3(x^P) + x c_3(x^P)], \end{aligned}$$

showing that (8) is satisfied with

$$\bar{a}(x) = a_4(x^P), \quad \bar{b}(x) = b_2(x^P) b_3(x^P) c_4(x^P) + x b_4(x^P),$$
$$\bar{c}(x) = c_4(x^P)\left[c_2(x^P) b_3(x^P) + x c_3(x^P)\right], \quad \bar{d}(x) = d_4(x^P).$$

It therefore suffices to prove that these four polynomials are $P$-linear. For the first and the fourth ones this is trivial. As for the two remaining polynomials, it is sufficient to prove that the three products

$$b_2(x) b_3(x) c_4(x), \quad c_2(x) b_3(x) c_4(x), \quad c_3(x) c_4(x) \tag{15}$$

are $P$-linear. This could easily be done for $P \neq 3$ (see the following Note 4), but the proof to be given here is valid for all $P$. We shall make repeated use of Proposition 16, without further reference.

By $(11_1)$, $(13_1)$, $(14_3)$, the (similar) direct decompositions

$$a_0(x) = a_2(x) \times b_2(x), \quad a_2(x) = a_3(x) \times b_3(x), \quad a_3(x) = a_4(x) \times c_4(x)$$

are valid, whence by substitution

$$a_0(x) = a_3(x) \times b_2(x) \times b_3(x) \times c_4(x) = a_3(x) \times b_2(x) b_3(x) c_4(x);$$

$(15_1)$ is therefore $P$-linear.

By $(11_2)$, $(13_4)$, $(14_2)$ we have

$$d_0(x) = c_2(x) \times d_2(x), \quad d_2(x) = b_3(x) \times d_3(x), \quad d_3(x) = c_4(x) \times d_4(x),$$

whence

$$d_0(x) = c_2(x) b_3(x) c_4(x) \times d_4(x),$$

implying the $P$-linearity of $(15_2)$.

Finally,

$$b_1(x) = a_3(x) \times c_3(x), \quad a_3(x) = a_4(x) \times c_4(x)$$

hold by $(13_3)$ and $(14_3)$, and consequently

$$b_1(x) = a_4(x) \times c_3(x) c_4(x),$$

which shows that $(15_3)$, too, is $P$-linear. Thus we have proved Lemma II.

NOTE 4. In

$$\max_i f_i^{\circ_P} \leqq \left( \prod_{i=1}^{k} f_i(x) \right)^{\circ_P} \leqq \sum_{i=1}^{k} f_i^{\circ_P} \qquad (f_i(x) \in \mathsf{F}[x];\ i = 1, \ldots, k)$$

the second $\leqq$ clearly holds for all $P$ and the first $\leqq$ holds, too, if the right-hand side is $\leqq P-1$. On the strength of this the last part of the proof for $P \neq 3$ can be replaced by the following abbreviated proof. From the second and third of the four equations that precede (15) we deduce that

$$\bar{b}^{\circ_P}, \ \bar{c}^{\circ_P} \leqq 3.$$

But we have proved that (8) is satisfied for the left-hand sides of the same equations, so that these inequalities must hold for $P \neq 3$ even with 1 instead of 3.

## § 22. Two group-theoretical propositions

As a last aid, we prove two propositions of some interest also in themselves for arbitrary groups. We shall need these propositions only for finite abelian groups but the greater generality will not make the proof any longer.

Let $(K)$ denote the subgroup of a given group, generated by a non-void complex $K$ of this group.

PROPOSITION 18. *If $K$ and $L \ni 1$ are complexes of a group $G$ such that the inclusion*

$$\gamma K \cap L \neq \emptyset \Rightarrow \gamma K \supseteqq L \tag{1}$$

*holds for the elements $\gamma \in G$ then $K$ is the union of left classes* mod $(L)$.

To prove this, consider arbitrary elements

$$\varkappa \in K, \quad \lambda \in L$$

and put

$$\gamma = \lambda \varkappa^{-1}.$$

We obtain $\lambda \in \gamma K$, and so, by (1), $\gamma K \supseteqq L$, $K \supseteqq \gamma^{-1} L \supseteqq \varkappa \lambda^{-1} L$. Thus

$$K \supseteqq K L^{-1} L$$

holds. Since $1 \in L$, this implies $K \supseteqq KL$, $KL^{-1}$, whence Proposition 18 follows.

A complex $K$ of a group $G$ is said to be *left primitive* if no subgroup $U \neq 1$ of $G$ exists for which $K$ is the union of left classes mod $U$. (In the commutative case, as well as in modules we use the term *primitive*, instead of left primitive.)

PROPOSITION 19. *If $K$ is a left primitive finite complex of a group $G$, then*

$$\bigcap_{\varkappa \in K} \varkappa^{-1} K = 1. \tag{2}$$

To prove this, put

$$D = \bigcap_{\varkappa \in K} \varkappa^{-1} K. \tag{3}$$

We shall show that for every $\gamma \in G$ the inclusion

$$\gamma^{-1} K \cap D \neq \emptyset \Rightarrow \gamma \in K \tag{4}$$

holds.

For the left-hand side of (4) implies the existence of a $\delta \in D$ with $\delta \in \gamma^{-1} K$, i.e.

$$\gamma \delta \in K.$$

On the other hand, from (3), $\delta \in \varkappa^{-1}K$, i.e.

$$\varkappa\delta \in K$$

for all $\varkappa \in K$. Since these $\varkappa\delta$ are different, it follows from these two results that $\gamma\delta$ agrees with a $\varkappa\delta$, i.e. $\gamma$ with $\varkappa$. This proves (4).

By (3) the validity of $\gamma^{-1}K \supseteqq D$ follows from $\gamma \in K$, and consequently (4) can be replaced by the stronger relation

$$\gamma^{-1}K \cap D \neq \varnothing \Rightarrow \gamma^{-1}K \supseteqq D.$$

As $\gamma^{-1}$ can here be any element of $G$, it follows from Proposition 18 (as $1 \in D$) that $K$ is the union of left classes mod $(D)$. Since, however, $K$ is left primitive, $D=1$ must hold. Thus we have proved Proposition 19.

## § 23. Transformation of Problem III. A necessary condition for the solutions

We begin our investigation of Problem III with an (equivalent) rewording, however, not yet applying the results of the three preceding paragraphs at this stage. Here $p=2$ is also allowed.

Let $P$ and $\varphi(x)$ be defined by

$$P \underset{1}{|} q \tag{1}$$

and

$$\varphi(x) \underset{1}{\in} \mathsf{F}[x], \quad \varphi(x) | x^P - x, \quad \frac{p-1}{2} \leqq \varphi^\circ \leqq P-1. \tag{2}$$

We must repeatedly draw the reader's attention to the equivalence of (1) with the fact that $q$ is a power of $P$ and that $O(\mathsf{F}_P)=P$ holds.

Let $f(x)$ denote an arbitrary stem polynomial (belonging to $P$). Then we know that $f(x)$ is defined by the four conditions

$$f(x) \underset{1}{\in} \mathsf{F}[x], \quad f^\circ = \frac{q-1}{P-1}, \quad f^{\circ_P} = 1, \quad x^q - x | f(x)^P - f(x). \tag{3}$$

We point out once more that these $f(x)$ are completely known (from §§ 3, 4).

The solutions of Problem III are then those $f(x)$ that satisfy the divisibility condition

$$\varphi(f(x)) | x^q - x. \tag{4}$$

More precisely, we say with the above terminology that an $f(x)$ with (3) and (4) is a solution of Problem III, belonging to the pair $P$, $\varphi(x)$.

For brevity we call every $\varphi(x)$ with properties (2) a *reference polynomial* for $P$. We naturally understand the validity of (1) and (2) to be included in the assertion "$f(x)$ is a solution of Problem III, belonging to $P$, $\varphi(x)$"; consequently $\varphi(x)$ is then a reference polynomial for $P$. It should be emphasized that (because of $(2_2)$) $(2_1)$ holds even with $\mathsf{F}_P$ in place of $\mathsf{F}$ and, moreover, that $\varphi(x)$ is fully reducible over $\mathsf{F}_P$.

We note that Problem III is clearly translation invariant (as the properties (3), (4) possess this property). It is even true that for a solution $f(x)$ of Problem III, belonging to the pair $P$, $\varphi(x)$, all translated $f(x+\xi)$ $(\xi \in \mathsf{F})$ belong to the same pair.

Since the condition (4) is now very complicated, it would hopeless to seek solutions in a direct way. A device that gives a "fortunate" transformation of (4) will see us through the difficulties also in this case.

To this end we shall use the component representation of $f(x)$ which, from $(3_3)$, can be written in the form

$$f(x) = f_0(x)^P + x f_1(x)^P, \tag{5}$$

so that after the introduction of the notation

$$q_1 = \frac{q}{P}, \tag{6}$$

by $(3_1)$, $(3_2)$, $(3_3)$ we have

$$f_0(x) \in \mathsf{F}[x], \quad f_1(x) \underset{1}{\in} \mathsf{F}[x]; \quad f_0^{\circ} \leqq f_1^{\circ} = \frac{q_1 - 1}{P-1}; \quad f_0^{\circ_P} \leqq f_1^{\circ_P} = 1. \tag{7}$$

We call $f_0(x), f_1(x)$ the *one-step constituents* of $f(x)$. (Also "two- and three-step constituents" will be defined later.) We notice that in the special case $P = p$ the one-step constituents have been defined already earlier (cf. § 18 (21) and (25)).

From (5) and (6) it follows that

$$f(x) \equiv \left(f_0(x) + x^{q_1} f_1(x)\right)^P \qquad (\text{mod } x^q - x).$$

From this and from $(3_4)$ we have the divisibility

$$x^q - x \,\big|\, \left(f(x) - f_0(x) - x^{q_1} f_1(x)\right)^P.$$

Since the left-hand side has no multiple factors, the exponent $P$ on the right-hand side can be cancelled. Then the new right-hand side is of degree $< q$, so that it must vanish. This means that

$$f(x) = f_0(x) + x^{q_1} f_1(x). \tag{8}$$

We say that the two representations (5) and (8) of the stem polynomials are *dual* to one another.

By (5), (6), (8) we have

$$f(x)^P - f(x) = (x^q - x) f_1(x)^P.$$

We transform this equation into

$$\frac{f(x)^P - f(x)}{f_1(x)^P \varphi(f(x))} = \frac{x^q - x}{\varphi(f(x))}.$$

It follows that the condition (4) can be replaced by

$$f_1(x)^P \varphi(f(x)) | f(x)^P - f(x).$$

With the aim of further transforming this divisibility, on the strength of $(2_2)$ we define the polynomial $\bar{\varphi}(x)$ by

$$\varphi(x)\bar{\varphi}(x) = x^P - x \tag{9}$$

and call it the *cofactor* of the reference polynomial (in $x^P - x$). It follows that

$$\varphi(f(x))\bar{\varphi}(f(x)) = f(x)^P - f(x)$$

and therefore that the above divisibility becomes

$$f_1(x)^P | \bar{\varphi}(f(x)).$$

Fortunately, this simplifies by (5) into

$$f_1(x)^P | \bar{\varphi}(f_0(x)^P).$$

Here, however, the $P$-th root can be extracted on both sides, whence

$$f_1(x) | \bar{\varphi}(f_0(x)). \tag{10}$$

Here we have used the fact that not only $\varphi(x)$ but also $\bar{\varphi}(x)$ lies in $\mathsf{F}_P[x]$, i.e. its coefficients are invariant under the automorphism $\xi \to \xi^P$. This proves the important

PROPOSITION 20. *If the validity of* (3) *is assumed, then the conditions* (4) *and* (10) *are equivalent to one another. Consequently, the solutions of Problem III, belonging to the pair $P$, $\varphi(x)$, are the $f(x)$ with the properties* (3) *and* (10).

On the whole (10) is still rather more complicated than (4), so that one could believe that no advantage is gained from Proposition 20. This is, however, not the case, as in a divisibility relation much more depends on the left-hand than on the right-hand side, and the former is much simpler in (10) than in (4). Anyhow, Proposition 20 constitutes merely the starting-point for further discussion, leading to a necessary condition, strong enough to enable us to deal successfully with Problem III.

NOTE 1. In general, the greater the ratio $\frac{h^\circ}{g^\circ}$ of the degrees $h^\circ$ and $g^\circ$, the greater the difficulty of dealing with the divisibility

$$g(x)|h(x) \quad (g(x), h(x) \in \mathsf{F}[x] \setminus \mathsf{F}).$$

For this reason this ratio may be called the *measure of difficulty* of the divisibility in question. (For $\frac{h^\circ}{g^\circ} < 1$ the divisibility does not hold, whereas for $\frac{h^\circ}{g^\circ} = 1$ it is equivalent to the "associatedness" of $g(x)$ and $h(x)$, so that only the case $\frac{h^\circ}{g^\circ} > 1$ presents any difficulty.) The measure of difficulty of (4) is evidently less than 2, whereas that of (10) can reach the number

$$\bar{\varphi}^\circ \left( = P - \varphi^\circ \leqq \frac{P+1}{2} \right);$$

in general it is therefore very great; however, (10) will guide us to another divisibility whose measure of difficulty surpasses 1 only by a small amount. The promised necessary condition will arise thus.

We shall use the linear factor decomposition

$$\bar{\varphi}(x) = \prod_{\bar{\varphi}(\varrho)=0} (x - \varrho) \tag{11}$$

where—as indicated—$\varrho$ has to run through the zeros of $\bar{\varphi}(x)$ (lying in $\mathsf{F}_P$) (We have used the fact that $\bar{\varphi}(x)$, being a divisor of $x^P - x$, is fully reducible over $\mathsf{F}_P$ and has no multiple zeros.) By (11), the divisibility (10) becomes

$$f_1(x) \Big| \prod_{\bar{\varphi}(\varrho)=0} (f_0(x) - \varrho).$$

This divisibility, after the introduction of the greatest common divisor

$$\Delta_\varrho(x) = (f_0(x) - \varrho, f_1(x)) \qquad (\varrho \in \mathsf{F};\ \bar{\varphi}(\varrho) = 0) \tag{12}$$

trivially becomes

$$f_1(x) \Big| \prod_{\bar{\varphi}(x)=0} \Delta_\varrho(x). \tag{13}$$

We have thus found, first of all, that Proposition 20 can be expressed in the following second form:

PROPOSITION 21. *Under assumption* (3) *the conditions* (4) *and* (13) *are equivalent to one another, and therefore the solutions of Problem III, belonging to the pair* $P$, $\varphi(x)$ *are the* $f(x)$ *with the properties* (3) *and* (13).

We interrupt our discussion to determine the linear solutions of Problem III as this case would disturb us later.

PROPOSITION 22. *A solution of Problem III is linear if and only if it belongs to* $P=q$ *(and to an arbitrary reference polynomial* $\varphi(x)$*); the solutions of Problem III, belonging to* $P=q$ *and to an arbitrary* $\varphi(x)$ *are (independently of* $\varphi(x)$*) all the* $x+\xi$ $(\xi\in \mathsf{F})$.

The first part of this proposition is a trivial conclusion from $(3_2)$. Since, moreover, for $P=q$ all the $f(x)$ with the property (3) are the $f(x) = x+\xi$ $(\xi\in\mathsf{F})$ and for these $f(x)$ (4) is also satisfied, as a consequence of $(2_2)$, Proposition 22 is proved.

(A part of this proposition asserts that all monic linear polynomials from $\mathsf{F}[x]$ are solutions of Problem III, where we are not concerned with the reference polynomial $\varphi(x)$.)

In the remainder of this section we shall no longer consider the linear solutions of Problem III, i.e. we shall henceforth assume that

$$P^2 \leqq q. \tag{14}$$

(In our later investigations, however, the linear solutions will constitute the starting point for a recursion procedure.)

From (14), the formations leading from $f(x)$ to $f_0(x)$, $f_1(x)$ can be repeated for the latter (in place of $f(x)$). By this we understand that, taking (7) into consideration, we put

$$f_j(x) = f_{0j}(x)^P + x f_{1j}(x)^P \qquad (j=0,1), \tag{15}$$

where

$$q_2 = \frac{q_1}{P}\left(=\frac{q}{P^2}\geqq 1\right), \tag{16}$$

and the conditions

$$f_{ij}(x)\in\mathsf{F}[x],\quad f_{11}(x)\underset{1}{\in}\mathsf{F}[x];\quad f_{ij}^{\circ}\leqq f_{11}^{\circ}=\frac{q_2-1}{P-1};\quad f_{ij}^{\circ P}\leqq f_{11}^{\circ P}=1 \tag{17}$$

$$(i,j=0,1)$$

are satisfied. It is clear that the four polynomials $f_{ij}(x)$ are thereby uniquely determined. They are called the *two-step constituents* of $f(x)$.

From (5) and (8) it follows that

$$f_0(x)^P + x f_1(x)^P = f_0(x) + x^{q_1} f_1(x).$$

On the right-hand side we substitute (15). Thus the equations

$$f_0(x)^P = f_{00}(x)^P + x^{q_1} f_{01}(x)^P, \quad f_1(x)^P = f_{10}(x)^P + x^{q_1} f_{11}(x)^P$$

are obtained by component comparison, whence, from (16), by taking the $P$-th root we get the equations

$$f_i(x) = f_{i0}(x) + x^{q_2} f_{i1}(x) \qquad (i = 0, 1) \tag{18}$$

which are *dual* to (15).

After substituting into (12) we have

$$\Delta_\varrho(x) = \big(f_{00}(x) - \varrho + x^{q_2} f_{01}(x), f_{10}(x) + x^{q_2} f_{11}(x)\big).$$

A decisive change takes place owing to the circumstance that Lemma 1 (p. 89) applies here. Thus we get the divisibility condition

$$\Delta_\varrho(x) \,\Bigg|\, \begin{vmatrix} f_{00}(x) - \varrho & f_{01}(x) \\ f_{10}(x) & f_{11}(x) \end{vmatrix} \tag{19}$$

which vill have important consequences.

We also introduce the notation

$$D(x) = \begin{vmatrix} f_{00}(x) & f_{01}(x) \\ f_{10}(x) & f_{11}(x) \end{vmatrix} \tag{20}$$

and call $D(x)$ the *determinant* (of the two-step constituents) of $f(x)$. Then (19) becomes

$$\Delta_\varrho(x) | D(x) - \varrho f_{11}(x)$$

and thus the required necessary condition

$$f_1(x) \Bigg| \prod_{\bar{\varphi}(\varrho)=0} \big(D(x) - \varrho f_{11}(x)\big) \tag{21}$$

follows from (13); after having done the multiplications and introduced the coefficient representation

$$\bar{\varphi}(x) = \sum_{i=0}^{\bar{\varphi}^\circ} \varkappa_i x^{\bar{\varphi}^\circ - i} \qquad (\varkappa_i \in \mathsf{F}_P), \tag{22}$$

this condition can be written in the second form

$$f_1(x) \Bigg| \sum_{i=0}^{\bar{\varphi}^\circ} \varkappa_i D(x)^{\bar{\varphi}^\circ - i} f_{11}(x)^i. \tag{23}$$

Thus we have proved

PROPOSITION 23. *For* $P^2 \leqq q$ (21) (*or* (23)) *follows from* (3) *and* (4). *Consequently,* (21) (*just like* (23)) *is a necessary condition for an* $f(x)$ (*satisfying* (3)) *to be a solution of Problem III, belonging to the pair* $P$, $\varphi(x)$.

This proposition will be of fundamental importance in our later investigations.

We now introduce the notation

$$D_\varrho(x) = D(x) - \varrho f_{11}(x) \left( = \begin{vmatrix} f_{00}(x) - \varrho & f_{01}(x) \\ f_{10}(x) & f_{11}(x) \end{vmatrix} \right) \qquad (\bar{\varphi}(\varrho) = 0) \qquad (24)$$

for the factors in (21) and call these determinants the *codeterminants* (of the two-step constituents) of $f(x)$. With their aid (21) can be expressed in the third form

$$f_1(x) \Big| \prod_{\bar{\varphi}(\varrho)=0} D_\varrho(x). \qquad (25)$$

NOTE 2. We are now faced with two essentially different cases, according to whether the right-hand side of the divisibility condition (21) is or is not equal to 0. The first case is characterized by the fact that one of the codeterminants $D_\varrho(x)$ vanishes; for this reason such a solution is said to be "degenerate", whereas the others are called "non-degenerate". (We shall later introduce other terminology for these concepts.) Accordingly, the investigations will be continued in two very different directions. It should be stressed that both cases are based on very strong assumptions. In the degenerate case this is clear; in this case Lemma II will apply very successfully. In the non-degenerate case the divisibility (21) (with a non-vanishing right-hand side) will render us good service, as their measure of difficulty (see the above Note 1) is not too great, being less than $1+\frac{1}{P}$, as is easily seen. (We mention this here only for the sake of orientation, as we shall no longer consider this measure of difficulty.) Please do not jump to any hasty conclusions, as in both cases we still have a very hard task ahead of us.

We now make some assertions which will come in useful later.

PROPOSITION 24. *If* $f(x)$ *is a solution of Problem III, belonging to the pair* $P$, $\varphi(x)$, *then* $f(x)+\xi$ *is a solution of this problem, belonging to the pair* $P$, $\varphi(x-\xi)$, *where* $\xi$ *may be any element of* $\mathsf{F}_P$,

Briefly, but, of course, somewhat imprecisely, this proposition means the parallel invariance of the solutions of Problem III.

COROLLARY. *All solutions of Problem III are* (*apart from parallel transformations*) *already covered by the case* $x \nmid \varphi(x)$.

(Of course, the case $f^\circ=1$, i.e. $P=q$ is an exception neither in Proposition 24 nor in the corollary.)

For, since the validity of (2), (3), (4) is not impaired by the simultaneous performance of the substitutions

$$f(x) \to f(x)+\xi, \quad \varphi(x) \to \varphi(x-\xi),$$

Proposition 24 holds. Applying the latter with a $\xi$ which is not a zero of $\varphi(-x)$ (from $(2_3)$ nothing prevents this from being done), the corollary follows.

We note that after substitution of (18) the equation

$$f(x) = f_{00}(x)+x^{q_2}f_{01}(x)+x^{q_1}f_{10}(x)+x^{q_1+q_2}f_{11}(x) \tag{26}$$

arises from (8). Trivially, this implies the following proposition.

PROPOSITION 25. *If $P^2 \leqq q$ the two-step constituents of a stem polynomial $f(x)$ (belonging to $P$), are completely characterized by* (26) *and*

$$f_{ij}(x) \in \mathsf{F}[x], \quad f_{ij}^\circ < q_2 \qquad (i,j=0,1). \tag{27}$$

We now examine the behaviour of the two-step constituents $f_{ij}(x)$ and of the determinant $D(x)$ under the translation $x \to x+\xi$ $(\xi \in \mathsf{F})$. We have

$$(x+\xi)^{q_i} = x^{q_i}+\xi^{q_i} \qquad (i=1,2)$$

and

$$(x+\xi)^{q_1+q_2} = x^{q_1+q_2}+\xi^{q_1}x^{q_2}+\xi^{q_2}x^{q_1}+\xi^{q_1+q_2}.$$

If (26) is transformed by $x \to x+\xi$, then, from Proposition 25, the substitution of these formulae leads to

PROPOSITION 26. *The two-step constituents of $f(x+\xi)$ $(\xi \in \mathsf{F})$ can be expressed in terms of those of $f(x)$, i.e. $f_{00}(x)$, $f_{01}(x)$, $f_{10}(x)$, $f_{11}(x)$, successively, as follows:*

$$f_{00}(x+\xi)+\xi^{q_2}f_{01}(x+\xi)+\xi^{q_1}f_{10}(x+\xi)+\xi^{q_1+q_2}f_{11}(x+\xi),$$
$$f_{01}(x+\xi)+\xi^{q_2}f_{11}(x+\xi),$$
$$f_{10}(x+\xi)+\xi^{q_1}f_{11}(x+\xi),$$
$$f_{11}(x+\xi).$$

*$f(x)$ and its determinant $D(x)$ are connected with one another in a way which is invariant under translation*, i.e. *the determinant of $f(x)+\xi$ is equal to $D(x+\xi)$.* (Of course, in this proposition $P^2 \leqq q$ is again assumed.)

The notations introduced here will be retained in what follows and we shall often apply the contents of this section without explicit reference.

## § 24. Non-primitive and primitive solutions of Problem III. Reduction of the former

The subsequent definition is of fundamental importance for what is to follow: A solution of Problem III, belonging to $P$, $\varphi(x)$, i.e. an $f(x)$ with

$$f(x) \underset{1}{\in} \mathsf{F}[x], \quad f^{\circ} = \frac{q-1}{P-1}, \quad f^{\circ P} = 1, \quad x^P - x | f(x)^P - f(x) \tag{1}$$

and

$$\varphi(f(x)) | x^q - x \tag{2}$$

is called *non-primitive* if ($P < q$ and) an equation

$$f(x) = g(x)^{\frac{Q-1}{P-1}} + \varrho \tag{3}$$

holds, where $Q$, $g(x)$ and $\varrho$ satisfy the conditions

$$P \underset{1}{|} Q \underset{1}{|} q, \quad P < Q \leqq q, \quad g(x) \in \mathsf{F}[x], \quad g^{\circ}{}_{\varrho} = 1, \quad \varrho \in \mathsf{F}^P \tag{4}$$

whereas in the opposite case the solution under consideration is called *primitive*. (Thus the linear solutions are primitive.) In the first case any $g(x)$ with the enumerated properties will be said to be a *successor* of $f(x)$, denoted by

$$f(x) \circ\!\!\rightarrow g(x).$$

It should be borne in mind that then at all times $f^{\circ} > g^{\circ}$ and that clearly also $Q$ and $P$ are already uniquely determined by $f(x)$ and $g(x)$. (No successor will be defined for the primitive solutions.) It will transpire, *inter alia*, that each successor must again be a solution of Problem III (this may, however, be either primitive or non-primitive). It should be pointed out that accordingly there is an (antireflexive and antisymmetric) binary relation in the set of all solutions of Problem III. It is "partially" transitive, too, whereby we understand that under certain circumstances (to be determined later) $f(x) \circ\!\!\rightarrow g(x)$ and $g(x) \circ\!\!\rightarrow h(x)$ together imply $f(x) \circ\!\!\rightarrow h(x)$.

NOTE 1. The reader might well have expected that the necessary condition formulated in Proposition 23 would be taken as the starting-point for our subsequent discussion. In fact, it will turn out that the non-primitive solutions of Problem III coincide with the degenerate ones (see § 23, Note 2) and the non-linear primitive solutions with the non-degenerate ones, but we shall obtain this result only as a result of a complicated argument (to be carried out in § 27). It has seemed advisable to bring the concept of the non-primitive solutions forward, as we shall apply it soon, in fact already in the next section, when we construct the "regular" solutions of Problem III. These will be of paramount importance, also in connection with our main problem, i.e. the marginal case of Problem II. We originally did the work in a different order, but the rearrangement used here seemed advisable for greater clarity.

NOTE 2. If for a $g(x) \in \mathsf{F}[x]$ conditions (3) and (4) are satisfied with the exception of the property of being monic, formulated in $(4_3)$, and $g^*(x)$ is the monic polynomial associated with $g(x)$, then conditions (3) and (4) are clearly completely satisfied for $g^*(x)$. Accordingly, in the above definition, the restriction that $g(x)$ be monic is inessential, but just therefore all the more desirable.

NOTE 3. Owing to Proposition 24, the condition (3) of non-primitivity means that one of the solutions of Problem III parallel to $f(x)$ is the $\frac{Q-1}{P-1}$-th power of a polynomial. Although this Note yields a conceptual simplification, we suitably continue to keep the original definition in mind.

We intend here to establish the simpler properties of the non-primitive solutions. We first have

PROPOSITION 27. *From* (1)—(4) *it follows:*

$$\varphi(\varrho) \neq 0 \qquad (i.e.\ \bar{\varphi}(\varrho)=0). \tag{5}$$

(This tallies with the notations used in § 23, where an (arbitrary) zero of the cofactor $\bar{\varphi}(x)$ was denoted by $\varrho$.)

For, (2) and (3) imply that

$$\varphi\left(g(x)^{\frac{Q-1}{P-1}}+\varrho\right) \Big| x^q - x. \tag{6}$$

(This divisibility will be used also later.) If now (5) is not true, i.e. $\varphi(\varrho)=0$, then (6) implies the divisibility

$$g(x)^{\frac{Q-1}{P-1}} \Big| x^q - x.$$

The right-hand side has no multiple divisors, while on the left-hand side the exponent is greater than 1, as a consequence of $(4_2)$. This is a contradiction, however, as $g(x)$ is not a constant. This proves Proposition 27.

PROPOSITION 28. *Under the assumption* $\varphi(x) \in \mathsf{F}[x]$ *and* $(4_1)$, $(4_5)$, (5)

$$\psi(x) = \varphi\left(x^{\frac{Q-1}{P-1}}+\varrho\right) \tag{7}$$

*is a reference polynomial for* $Q$ *if and only if* $\varphi(x)$ *is such a polynomial for* $P$.

It follows from (7) that the conditions $\varphi(x) \underset{1}{\in} \mathsf{F}[x]$ and $\psi(x) \underset{1}{\in} \mathsf{F}[x]$ are equivalent. Moreover, the same holds for the conditions

$$\frac{P-1}{2} \leqq \varphi^{\circ} \leqq P-1, \quad \frac{Q-1}{2} \leqq \psi^{\circ} \leqq Q-1.$$

Only the equivalence of the conditions

$$\varphi(x)|x^P - x, \quad \psi(x)|x^Q - x \tag{8}$$

remains to be proved.

Because of $(4_5)$, condition $(8_1)$ is equivalent to

$$\varphi(x+\varrho)|x^P - x. \tag{9}$$

From (5), the left-hand side of (9) is not divisible by $x$, and therefore (9) is equivalent to

$$\varphi(x+\varrho)|x^{P-1} - 1.$$

The latter divisibility can, however, be subjected to the transformation

$$x \to x^{\frac{Q-1}{P-1}},$$

which leads to the equivalent divisibility

$$\varphi\left(x^{\frac{Q-1}{P-1}} + \varrho\right)\Big| x^{Q-1} - 1.$$

Here again the left-hand side is not divisible by $x$, and so we get an equivalent divisibility when we replace the right-hand side by $x^Q - x$, which, applying (7), gives $(8_2)$. This proves Proposition 28.

We have already made the trivial remark that Problem III is translation invariant. The next proposition can be inferred in a similarly trivial manner from the definition.

PROPOSITION 29. *The property of the solutions of Problem III of being primitive (and consequently also that of being non-primitive) is translation invariant. More precisely,* $f(x+\xi) \multimap g(x+\xi)$ *follows from* $f(x) \multimap g(x)$ *for every* $\xi \in \mathsf{F}$.

We shall also be concerned with chains of the form

$$f(x) \multimap g(x) \multimap \ldots$$

called *successor chains*. As, then, $f^{\circ} > g^{\circ} > \ldots$, every successor chain is finite. Attention should be paid to the fact that the relation $\multimap$, in general, is not transitive.

To answer this question also indicated above, assume $f(x) \multimap g(x)$ and $g(x) \multimap h(x)$. Then we have (*inter alia*)

$$f(x) = g(x)^{\frac{Q-1}{P-1}} + a, \quad g(x) = h(x)^{\frac{R-1}{Q-1}} + \sigma \quad (\varrho \in \mathsf{F}_P,\ \sigma \in \mathsf{F}_Q)$$

and

$$P \underset{1}{|} Q \underset{1}{|} R \underset{1}{|} q, \quad P < Q < R < q, \quad h^{\circ_R} = 1.$$

As this implies that

$$f(x) = \left(h(x)^{\frac{R-1}{Q-1}} + \sigma\right)^{\frac{Q-1}{P-1}} + \varrho,$$

it is clear that

$$f(x) = h(x)^{\frac{R-1}{P-1}} + \tau$$

holds in the case $\sigma = 0$ for $\tau = \varrho$ and in the case $\sigma \neq 0$ for no $\tau \in \mathsf{F}_P$. Thus we have proved

PROPOSITION 30. *When $f(x) \multimap g(x) \multimap h(x)$, the relation $f(x) \multimap h(x)$ holds if and only if $g(x)$ is a power of $h(x)$.*

We now have the problem of surveying the set of all successors of a given $f(x)$. In this context we prove

PROPOSITION 31. *If $a(x)$, $b(x)$ are successors of $f(x)$, so that we may write*

$$f^{\circ} = \frac{q-1}{P-1}, \quad a^{\circ} = \frac{q-1}{A-1}, \quad b^{\circ} = \frac{q-1}{B-1} \quad (P \underset{1}{|} A \underset{1}{|} q,\ P \underset{1}{|} B \underset{1}{|} q), \tag{10}$$

*then $f(x)$ has a successor $c(x)$ such that*

$$c^{\circ} = \frac{q-1}{C-1}, \quad A \underset{1}{|} C \underset{1}{|} q, \quad B \underset{1}{|} C \underset{1}{|} q, \quad a(x) = c(x)^{\frac{C-1}{A-1}}, \quad b(x) = c(x)^{\frac{C-1}{B-1}} \tag{11}$$

*(whence, of course, it follows that $a(x) \multimap c(x)$ and $b(x) \multimap c(x)$).*

COROLLARY. *Every non-primitive solution $f(x)$ of Problem III has a (clearly unique) successor $g(x)$ such that all the successors of $f(x)$ are given by*

$$g(x)^{\frac{Q-1}{X-1}} \qquad (P \underset{1}{|} X \underset{1}{|} Q;\ P < X \leqq Q) \tag{12}$$

*(with a variable $X$), where $P$ and $Q$ are determined from the formulae*

$$f^{\circ} = \frac{q-1}{P-1}, \quad g^{\circ} = \frac{q-1}{Q-1}. \tag{13}$$

According to our assumption, we have

$$f(x) = a(x)^{\frac{A-1}{P-1}} + \alpha, \quad f(x) = b(x)^{\frac{B-1}{P-1}} + \beta \qquad (\alpha, \beta \in \mathsf{F}_P). \tag{14}$$

Thus

$$a(x)^r + \alpha = b(x)^s + \beta, \tag{15}$$

where, for the sake of brevity, we have put

$$r = \frac{A-1}{P-1}, \quad s = \frac{B-1}{P-1}. \tag{16}$$

First of all we prove that

$$\alpha = \beta. \tag{17}$$

If this is false, then, by (15), it follows that

$$\big(a(x),\ b(x)\big) = 1.$$

Moreover, as by (16) it is also true that

$$r \equiv s \equiv 1 \qquad (\text{mod}\, p),$$

it follows from (15) that

$$a(x)^{r-1} a'(x) = b(x)^{s-1} b'(x),$$

and therefore also that

$$a(x)^{r-1} | b'(x), \quad b(x)^{s-1} | a'(x).$$

As, however, according to our assumption we have $a^{\circ}, b^{\circ} < f^{\circ}$, from (10) we have that $P < A, B$, and thus by (16) it follows that $r, s \geqq 2$. This gives

$$a(x) | b'(x), \quad b(x) | a'(x).$$

This, however, is a contradiction, as the right-hand sides are not 0. This proves (17).

On this account (15) simplifies into

$$a(x)^{\frac{A-1}{P-1}} = b(x)^{\frac{B-1}{P-1}}. \tag{18}$$

We take (uniquely determined) natural numbers $d$, $u$, $v$ with

$$A = P^{dv}, \quad B = P^{du} \tag{19}$$

and

$$(u, v) = 1; \tag{20}$$

furthermore, we put

$$C = P^{duv}, \tag{21}$$

whence

$$C = A^u = B^v \tag{22}$$

follows.

Now putting $(m = (\mathsf{F}|\mathsf{F}_P)^\circ$, i.e.)

$$P^m = q, \tag{23}$$

it will be seen that the validity of

$$du,\ dv | m$$

follows from

$$A,\ B \underset{1}{|} q$$

and (19). In view of (20) this implies

$$duv | m. \tag{24}$$

According to (21), (22), (23) and (24), the relations $(11_2)$, $(11_3)$ are satisfied; it suffices therefore to prove the existence of a $c(x)$ with $f(x) \multimap c(x)$, as well as $(11_1)$, $(11_4)$, $(11_5)$.

As $a(x)$ and $b(x)$ are monic, (18) implies the existence of a $c(x) \in \mathsf{F}[x]$ with

$$a(x) = c(x)^{\overline{\tfrac{s}{(r,\,s)}}}, \qquad b(x) = c(x)^{\overline{\tfrac{r}{(r,\,s)}}}.$$

Accordingly we have by (16)

$$a(x) = c(x)^{\overline{\tfrac{B-1}{(A-1,\,B-1)}}} \qquad b(x) = c(x)^{\overline{\tfrac{A-1}{(A-1,\,B-1)}}}. \tag{25}$$

From $(10_2)$ and $(25_1)$ we infer that

$$c^\circ = \frac{q-1}{[A-1,\ B-1]}, \tag{26}$$

where [ ] denotes the least common multiple.

Now by (19) and (20)

$$[A-1,\ B-1] = P^{duv} - 1$$

and therefore by (21)

$$[A-1,\ B-1] = C-1, \quad (A-1,\ B-1) = \frac{(A-1)(B-1)}{C-1}.$$

Thus (25) and (26) become $(11_4)$, $(11_5)$ and $(11_1)$. We have thus proved Proposition 31.

To prove the corollary let $g(x)$ be a successor of $f(x)$ of minimal degree. Here we may assume the validity of (13) and of

$$P \underset{1}{|} Q \underset{1}{|} q, \quad g(x) \underset{1}{\in} \mathsf{F}[x], \quad g^\circ \varrho = 1 \quad f(x) = g(x)^{\overline{\tfrac{Q-1}{P-1}}} + \varrho, \quad \varrho \in \mathsf{F}_P.$$

Denoting the polynomial (12) by $h(x)$ (which has nothing to do with the above $h(x)$), it is clear that

$$f(x) = h(x)^{\overline{\frac{X-1}{P-1}}} + \varrho, \quad h(x) \underset{1}{\in} \mathsf{F}[x], \quad h^{\circ x} = 1$$

using the fact that

$$h(x) = g(x) g(x)^{X} \ldots g(x)^{\overline{\frac{Q}{X}}};$$

and thus $h(x)$ really is (as $P \underset{1}{|} X \underset{1}{|} q$ and $P < X$) a successor of $f(x)$.

On the other hand, consider an arbitrary successor $a(x)$ of $f(x)$ for which we are allowed (as in (10)) to assume that

$$a^{\circ} = \frac{q-1}{A-1}, \quad P \underset{1}{|} A \underset{1}{|} q.$$

The application of Proposition 31 with $g(x)$ in place of $b(x)$ guarantees the existence of a successor $c(x)$ of $f(x)$ for which (*inter alia*)

$$c^{\circ} = \frac{q-1}{C-1}, \quad a(x) = c(x)^{\overline{\frac{C-1}{A-1}}}, \quad g(x) = c(x)^{\overline{\frac{C-1}{Q-1}}}$$

holds. Since, however, $g^{\circ}$ is minimal, $c^{\circ} \geqq g^{\circ}$ and consequently $C = Q$, $c(x) = g(x)$ must hold, whence

$$a(x) = g(x)^{\overline{\frac{Q-1}{A-1}}}$$

follows. This means that $a(x)$ appears among the polynomials (12), which proves the corollary.

We close this paragraph with the promised important

THEOREM 9. (Reduction theorem for the non-primitive solutions of Problem III.) *Every successor $g(x)$ of a (non-primitive) solution $f(x)$ of Problem III is again a solution of this problem; furthermore, if $f(x)$ belongs to the pair*

$$P, \varphi(x) \tag{27}$$

*and (because of the assumption)*

$$f(x) = g(x)^{\overline{\frac{Q-1}{P-1}}} + \varrho, \quad P \underset{1}{|} Q \underset{1}{|} q, \quad P < Q, \qquad \varrho \in \mathsf{F}_P \tag{28}$$

*holds, then $g(x)$ belongs to the pair*

$$Q, \varphi\left(x^{\overline{\frac{Q-1}{P-1}}} + \varrho\right). \tag{29}$$

*Conversely, if* (28) *holds and for a suitable* $\varphi(x) \underset{1}{\in} \mathsf{F}[x]$ *the polynomial* $g(x)$ *is a solution of Problem III, belonging to the pair* (29), *then* $f(x)$ *is a (non-primitive) solution of this problem, belonging to the pair* (27), *with the successor* $g(x)$.

In a few words, the non-primitive solutions of Problem III of given degree are reduced, by this theorem, to solutions of lower degree.

To prove the first half of the theorem, we first show that $g(x)$ is a stem polynomial for $Q$, i.e. that

$$g(x) \underset{1}{\in} \mathsf{F}[x], \quad g^{\circ} = \frac{q-1}{Q-1}, \quad g^{\circ}{}_{\varrho} = 1, \quad x^q - x | g(x)^Q - g(x) \tag{30}$$

holds.

Since $f(x) \circ\!\!\to g(x)$, $(30_1)$ is valid.

By the assumption

$$f(x) \underset{1}{\in} \mathsf{F}[x], \quad f^{\circ} = \frac{q-1}{P-1}, \quad f^{\circ}{}_{P} = 1, \quad x^q - x | f(x)^P - f(x). \tag{31}$$

As a consequence of (28) and $(31_2)$, we have $(30_2)$.

Because of $f(x) \circ\!\!\to g(x)$ and $(30_2)$, $(30_3)$ holds.

After the substitution of $(28_1)$, the divisibility $(31_4)$, using $(28_4)$, becomes

$$x^q - x | g(x)^{P\frac{Q-1}{P-1}} - g(x)^{\frac{Q-1}{P-1}}. \tag{32}$$

As the left-hand side has no multiple factors, the right-hand side can be divided by

$$g(x)^{\frac{Q-1}{P-1}-1}, \tag{33}$$

which gives us just $(30_4)$. Thus (30) is proved.

By reason of Proposition 27 we have

$$\varphi(\varrho) \neq 0 \tag{34}$$

and therefore by virtue of Proposition 28

$$\psi(x) = \varphi\left(x^{\frac{Q-1}{P-1}} + \varrho\right) \tag{35}$$

is a reference polynomial for $Q$.

Finally, the assumption implies that

$$\varphi(f(x)) | x^q - x. \tag{36}$$

Substituting $(28_1)$, we obtain

$$\varphi\left(g(x)^{\frac{Q-1}{P-1}} + \varrho\right) \Big| x^q - x. \tag{37}$$

By (30), (35), and (37), $g(x)$ is in fact a solution of Problem III, belonging to the pair $Q$, $\psi(x)$. This completes the proof of the first part of Theorem 9.

To prove the second part note that (30) and (37) now follow from our assumption. We must first show that $f(x)$ is a stem polynomial for $P$, i.e. that (31) holds.

$(28_1)$ and $(30_1)$ imply $(31_1)$.

$(28_1)$ and $(30_2)$ imply $(31_2)$.

Let us transform $(28_1)$ (using $(28_2)$) into

$$f(x) = g(x)g(x)^P \ldots g(x)^{\frac{Q}{P}} + \varrho.$$

Hence, and from $(30_3)$, $(31_3)$ follows.

If we multiply the right-hand side of $(30_4)$ by the power (33), we obtain the divisibility (32). Adding to its right-hand side the difference $\varrho^P - \varrho$, which vanishes by $(28_4)$, it is precisely $(31_4)$ that results if we apply $(28_1)$. Thus (31) is proved.

From $(28_3)$, it follows that the exponent on the left-hand side of (37) is greater than 1. Consequently if $\varphi(\varrho)=0$ the left-hand side would have multiple divisors. As this is impossible, (34) must hold. It follows from this and from Proposition 28 that $\varphi(x)$ is a reference polynomial for $P$. Finally, since (36) follows from $(28_1)$ and (37), Theorem 9 is proved.

NOTE 4. For each solution $f(x)$ of Problem III let $L(f)$ denote the set consisting of $f(x)$ and its successors. It follows from Theorem 9, the "if" part of Proposition 30 and Proposition 31 that the relation $\circ\!\!\rightarrow$ is transitive in $L(f)$ and we see that $L(f)$ forms a lattice.

NOTE 5. For each solution $f(x)$ (belonging to a pair $P$, $\varphi(x)$) of Problem III let $T(f)$ denote the set of all $f(x+\xi)$ $(\xi \in \mathsf{F})$. By Proposition 29 the different $T(f)$ form a partition of the set of all solutions of Problem III; for this reason we call the $T(f)$ *translation classes* (of the set) of the solutions of Problem III. It can also be inferred from Proposition 29 that every class $T(f)$ contains either only non-primitive or only primitive solutions.

NOTE 6. Let

$$P \underset{1}{|} Q \underset{1}{|} q, \quad P < Q (\leqq q)$$

and let us consider two polynomials $u(x), v(x) \in \mathsf{F}[x]$ with

$$u(x) = v(x)^{\frac{Q-1}{P-1}}.$$

Then by virtue of a simple reasoning applied above the inclusion

$$v^{\circ_Q} = 1 \Rightarrow u^{\circ_P} = 1$$

holds also in this generality. On the other hand, the converse

$$u^{\circ_P}=1 \Rightarrow v^{\circ_Q}=1$$

is false, as is shown by the example

$$P^2 = Q = q, \quad u(x) = (x+1)^{P^3+1}, \quad v(x) = (x+1)^{P^2-P+1},$$

where $u^{\circ_P}=1$ (even $u^{\circ_{P^3}}=1$), while $v^{\circ_Q} = P^2-P+1 > 1$. (Also $v^{\circ_P} = P-1$, so that even $v^{\circ_P}>1$ holds, except for $P=2$, $Q=q=4$.) It seems probable therefore that, in the definition of a successor, condition ($4_4$) is independent of the remaining conditions. This question is not of any relevance here, which is why we have not dealt with it.

## § 25. The regular solutions of Problem III

We now investigate the successor chains in more detail. This will lead us to the important "regular" solutions of Problem III. It is convenient to write a successor chain (of $k+1$ terms) in the form

$$f_{(0)}(x) \circ\!\!\rightarrow f_{(1)}(x) \circ\!\!\rightarrow \cdots \circ\!\!\rightarrow f_{(k)}(x). \tag{1}$$

According to the definition of a successor, the terms must here be, with at most the exception of the last term, non-primitive solutions of Problem III; moreover, by Theorem 9 also the last term $f_{(k)}(x)$ is a (non-primitive or primitive) solution of this problem. We shall also allow the case $k=0$ and then identify the chain (1) with an arbitrary solution of Problem III. We shall mainly be interested in the question of how to calculate the initial term from the last term in (1).

From the definition of the successor, (1) implies the existence of numbers $P_0, \ldots, P_k$ with

$$P_0 \underset{1}{|} P_1 \underset{1}{|} \cdots \underset{1}{|} P_k \underset{1}{|} q, \qquad P_0 < P_1 < \cdots < P_k \leqq q \tag{2}$$

and of reference polynomials

$$\varphi_i(x) \underset{1}{\in} \mathsf{F}_{P_i}[x] \qquad \left(\varphi_i(x) | x^{P_i} - x, \ \frac{P_i - 1}{2} \leqq \varphi_i \leqq P_i - 1; \ i = 0, \ldots, k\right) \tag{3}$$

such that $f_{(i)}(x)$ belongs to the pair $P_i$, $\varphi_i(x)$; moreover, the existence of elements

$$\varrho_i \in \mathsf{F}_{P_i} \qquad (i = 0, \ldots, k-1) \tag{4}$$

follows, such that after the introduction of the notation

$$e_i(x) = x^{\frac{P_{i+1}-1}{P_i-1}} + \varrho_i \qquad (i = 0, \ldots, k-1) \tag{5}$$

the equations

$$f_i(x) = e_i(f_{(i+1)}(x)) \qquad (i = 0, \ldots, k-1) \tag{6}$$

hold. By Propositions 27, 28, the conditions

$$\varphi_i(\varrho_i) \neq 0, \qquad \varphi_{i+1}(x) = \varphi_i(e_i(x)) \qquad (i = 0, \ldots, k-1) \tag{7}$$

must be satisfied here. The result of this consideration can be summarized in

PROPOSITION 32. *For any successor chain* (1)

$$P_0, \ldots, P_k; \quad \varphi_0(x), \ldots, \varphi_k(x); \quad \varrho_0, \ldots, \varrho_{k-1}$$

*can be chosen such that* (2)—(7) *are satisfied and* $f_{(i)}(x)$ *is a solution of Problem III, belonging to the pair* $P_i, \varphi_i(x)$ $(i=0, \ldots, k)$.

SUPPLEMENT. *From* (4), (5), (7) *and from the validity of* (3) *for* $i=0$ *its validity follows in general.*

The supplement follows from Proposition 28 by repeated application. (Similarly by Proposition 28, the supplement is, more generally, valid for every fixed $i=0, \ldots, k$, but we shall not need this degree of generality.)

We prove

PROPOSITION 33. *Under assumptions* (2)—(7) *the* $k+1$ *propositions:*

$$f_{(i)}(x) \text{ is a solution of Problem III, belonging to } P_{(i)}, \ \varphi_i(x) \quad (i=0, \ldots, k) \tag{8}$$

*are equivalent to one another.*

SUPPLEMENT. *Under assumptions* (2)—(8) *the chain* (1) *exists.*

To prove the proposition, let us assume (8) for a fixed $i$. It suffices to prove the two partial assertions that (8) is satisfied for $i+1$ in the case $i<k$ and for $i-1$ in the case $i>0$ (in place of $i$).

Suppose, first of all, that $i<k$. We put

$$f(x)=f_{(i)}(x), \quad g(x)=f_{(i+1)}(x), \quad P=P_i, \quad Q=P_{i+1}, \quad \varrho=\varrho_i, \quad \varphi(x)=\varphi_i(x).$$

By (8) and a part of (2)—(7) all assumptions of the first half of Theorem 9 are satisfied, and consequently $(g(x)=)f_{(i+1)}(x)$ is a solution of Problem III, belonging to the pair

$$(Q=)P_{i+1},\quad \left(\varphi\left(x^{\frac{Q-1}{P-1}}\right)+\varrho=\varphi_i\left(x^{\frac{P_{i+1}-1}{P_i-1}}+\varrho_i\right)=\right)\varphi_{i+1}(x).$$

Therefore (8) is satisfied for $i+1$.

Now let $i>0$. Put

$$f(x)=f_{(i-1)}(x),\quad g(x)=f_{(i)}(x),\quad P=P_{i-1},\quad Q=P_i,\quad \varrho=\varrho_{i-1},\quad \varphi(x)=\varphi_{i-1}(x).$$

First of all, it again follows from a part of (2)—(7) that

$$\varphi\left(x^{\frac{Q-1}{P-1}}+\varrho\right)=\varphi_{i-1}\left(x^{\frac{P_i-1}{P_{i-1}-1}}+\varrho_{i-1}\right)=\varphi_{i-1}(e_{i-1}(x))=\varphi_i(x),$$

and consequently that the assumptions of the second half of Theorem 9 are satisfied; so $(f(x)=)=f_{(i-1)}(x)$ is a solution of Problem III, belonging to the pair $(P=)P_{i-1}$, $(\varphi(x)=)\varphi_{i-1}(x)$. This proves Proposition 33. The supplement is trivial.

A successor chain (1) is said to be *abridgeable* if there exists an $i$ with $0<i<k$ such that the chain

$$f_{(0)}(x)\multimap f((_{(i-1)}(x)\multimap{}_{(i+1)}(x)\multimap\cdots\multimap f_{(k)}(x),$$

yielded by the cancelling of the $i+1$-th term, is also a correct successor chain. Since for this the validity of

$$f_{(i-1)}(x)\multimap f_{(i+1)}(x)$$

is necessary and sufficient, Proposition 30 implies the following

PROPOSITION 34. *A successor chain* (1) *is abridgeable if and only if no term* $f_{(i)}(x)$ $(0<i<k)$ *is a power of its successor* $f_{(i+1)}(x)$.

We now introduce the following definition: A solution $f(x)$ of Problem III is called *regular* if there exists a successor chain with the initial term $f(x)$ and with a linear last term. (In this definition we are, of course, allowed to confine ourselves to non-abridgeable successor chains.) This definition is to be understood so as to include, in particular, the linear solutions $x+\xi$ $(\xi\in\mathsf{F})$ among the regular ones.

With the aid of Proposition 33 it is easy to construct all the regular solutions of Problem III. For this purpose we shall restrict ourselves to successor chains (1) with a linear last term. According to Proposition 22, this is carried out by putting

$$P_k=q \tag{9}$$

and (independently of $\varphi_k(x)$)

$$f_{(k)}(x) = x + \varrho_k \qquad (\varrho_k \in \mathsf{F}). \tag{10}$$

(We have denoted the constant term by $\varrho_k$, in order to conform with the notations of (4). The effect of this will soon be clear.) Consequently, if (10) is substituted into (6), then (8) holds for all the $f_{(i)}(x)$ ($i=0, \dots, k$), originating in this way. After solving the recursion (6) the expressions of the latter are found to be

$$f_{(i)}(x) = e_i(\dots(e_{k-1}(x+\varrho_k)\dots) \qquad (i = 0, \dots, k). \tag{11}$$

We see that of these the case $i=0$ will give all the regular solutions of Problem III, and the same holds even if we confine ourselves to non-abridgeable successor chains (1). From Proposition 34, this means that for the $i=1, \dots, k-1$ the left-hand side of (6) is not a power of the right-hand side, i.e. (cf. (5)) all the $\varrho_1, \dots, \varrho_{k-1}$ are chosen to be different from 0.

Before formulating the result definitively, we intend to give a more explicit form of conditions $(7_1)$. Solving the recursion $(7_2)$ we obtain

$$\varphi_i(x) = \varphi_0(e_0(\dots(e_{i-1}(x))\dots)),$$

i.e., by (5),

$$\varphi_i(x) = \varphi_0\Big(\Big(\dots\Big(x^{\frac{P_i-1}{P_{i-1}-1}} + \varrho_{i-1}\Big)\dots\Big)^{\frac{P_1-1}{P_0-1}} + \varrho_0\Big) \qquad (i = 0, \dots, k).$$

Consequently, using the formula

$$\xi^{\frac{P_{i+1}-1}{P_i-1}} = N_{\mathsf{F}_{P_{i+1}}|\mathsf{F}_{P_i}}(\xi) \qquad (\xi \in \mathsf{F}_{P_{i+1}};\ i = 0, \dots, k-1)$$

$(7_1)$ becomes

$$\varphi_0(N_{\mathsf{F}_{P_1}|\mathsf{F}_{P_0}}(\dots(N_{\mathsf{F}_P|\mathsf{F}_{P_{i-1}}}(\varrho_i)+\varrho_{i-1})\dots)+\varrho_0) \neq 0 \qquad (i = 0, \dots, k-1).$$

In the outer brackets we find an element of $\mathsf{F}_{P_0}$ (dependent on $i$) which we shall, for the time being, denote by $\omega$, so that (12) takes the form $\varphi_0(\omega)=0$. However, this inequality is (because $\omega \in \mathsf{F}_{P_0}$) equivalent to the fact that $\omega$ is a zero of the cofactor

$$\varphi_0(x) = \varphi_0(x)^{-1}(x^{P_0} - x)$$

of $\varphi_0(x)$.

Writing $f(x)$ and $\varphi(x)$ in place of $f_0(x)$ and $\varphi_0(x)$, respectively, and taking into account the supplement to Proposition 32 we now summarize the result in

THEOREM 10. (Determination of the regular solutions of Problem III.) *For arbitrary $P$, $\varphi(x)$ with*

$$P|q, \quad \varphi(x) \in \mathsf{F}[x], \quad \varphi(x)|x^P - x, \quad \frac{P-1}{2} \leqq \varphi^{\circ} \leqq P-1 \tag{13}$$

*the formula*

$$f(x) = \left(\dots\left((x+\varrho_k)^{\frac{p_k-1}{p_{k-1}-1}}+\varrho_{k-1}\right)\dots\right)^{\frac{p_1-1}{p_0-1}}+\varrho_0 \tag{14}$$

*represents all regular solutions of Problem III belonging to the pair* $P, \varphi(x)$, *where*

$$\underset{1}{P_0}|\underset{1}{P_1}|\cdots|\underset{1}{P_k}|\underset{1}{q}, \quad P=P_0<P_1<\cdots<P_k=q \tag{15}$$

*holds and elements*

$$\varrho_i \in \mathsf{F}_{P_i} \qquad (i=0, \dots, k) \tag{16}$$

*are admitted chosen in such a way that all the relative norms*

$$N_{\mathsf{F}_{P_1}|\mathsf{F}_{P_0}}(\cdots(N_{\mathsf{F}_{P_i}|\mathsf{F}_{P_{i-1}}}(\varrho_i)+\varrho_{i-1})\cdots)+\varrho_0 \qquad (i=0, \dots, k-1) \tag{17}$$

*are zeros of the cofactor* $\bar{\varphi}(x)\left(=\varphi(x)^{-1}(x^P-x)\right)$ of $\varphi(x)$.

SUPPLEMENT. *All that has just been said holds also under the restriction*

$$\varrho_1, \dots, \varrho_{k-1} \neq 0 \tag{18}$$

*and the representation* (14) *of the regular solutions is then unique.*

It is only the assertion of uniqueness in the supplement, that remains to be proved. We write the right-hand side of (14) as

$$f(x; P_0, \dots, P_k; \varrho_0, \dots, \varrho_k).$$

It is sufficient to prove that from the assumption

$$\underset{1}{P_0}|\underset{1}{P_1}|\cdots|\underset{1}{P_k}|\underset{1}{q}, \quad P_0<P_1<\cdots<P_k, \tag{19}$$

$$f(x; P_0, \dots, P_k; \varrho_0, \dots, \varrho_k) = f(x; P_0, \dots, P_k; \sigma_0, \dots, \sigma_k), \tag{20}$$

$$\varrho_0, \dots, \varrho_k, \sigma_0, \dots, \sigma_k \in \mathsf{F}, \tag{21}$$

$$\varrho_1, \dots, \varrho_{k-1}, \sigma_1, \dots, \sigma_{k-1} \neq 0 \tag{22}$$

the equations

$$\varrho_i = \sigma_i \qquad (i=0, \dots, k) \tag{23}$$

follow.

When $k=0$ the assertion is true. When $k \geqq 1$ we suppose its validity for the smaller values of $k$. By applying the translation $x \to x-\sigma_k$, equation (20) becomes

$$f(x;\ P_0, \dots, P_k;\ \varrho_0, \dots, \varrho_{k-1}, \varrho_k-\sigma_k) = f(x;\ P_0, \dots, P_k;\ \sigma_0, \dots, \sigma_{k-1}, 0). \tag{24}$$

The right-hand side lies in

$$\mathsf{F}\left[x^{\frac{p_k-1}{p_{k-1}-1}}\right]$$

and consequently it has a sinking >1. The sinking of the left-hand side is, when $\varrho_k-\sigma_k \neq 0$, clearly equal to 1, and therefore $\varrho_k-\sigma_k = 0$, i.e.

$$\varrho_k=\sigma_k.$$

Then the substitution

$$x^{\frac{p_k-1}{p_{k-1}-1}}\to x$$

changes (24) into

$$f(x;\ P_0,\ \dots,\ P_{k-1};\ \varrho_0,\ \dots,\ \varrho_{k-1}) = f(x;\ P_0,\ \dots,\ P_{k-1},\ \sigma_0,\ \dots,\ \sigma_{k-1}),$$

whence by induction the validity of (23) follows for $i=0, \dots k-1$. Hence the last assertion and thus also the supplement are proved.

NOTE. As (2) and (4) imply

$$\varrho_i \in \mathsf{F}^{\frac{q-1}{p_i-1}} \subseteq \mathsf{F}^{\frac{p_{i+1}-1}{p_i-1}},$$

(5) is an Euler binomial when $\varrho_i \neq 0$. Accordingly it follows from Theorem 10 and the supplement that in the general regular solutions of Problem III arise by iteration from Euler polynomials and a subsequent translation. Only $\varrho_0=0$, $k>0$ is an exception, for which it is necessary to raise to some power at the end.

## § 26. Explicit determination of the regular solutions of Problem III

As a preparation for this question we consider two subfields

$$\mathsf{F}_P \subset \mathsf{F}_Q$$

of $\mathsf{F}$. This implies the validity of

$$P \underset{1}{|} Q \underset{1}{|} q, \quad P<Q\leqq q.$$

As is well-known, the relative norms

$$N_{\mathsf{F}_Q|\mathsf{F}_P}(\xi)=\xi^{\frac{Q-1}{P-1}}$$

of the elements $\xi \in \mathsf{F}_Q$ run through all elements of $\mathsf{F}_P$. For an $\alpha \in \mathsf{F}_Q$ we call the set of the elements $\xi \in \mathsf{F}_Q$ with

$$N_{\mathsf{F}_Q|\mathsf{F}_P}(\xi)=\alpha$$

the $(\mathsf{F}_Q, \mathsf{F}_P)$-*norm class* of $\alpha$. In the case $\alpha=0$ this norm class consists of the single element 0, in the case $\alpha\neq 0$, however, of $\dfrac{Q-1}{P-1}$ elements; morever, it is clear that its elements are all the values of the $\dfrac{Q-1}{P-1}$-th root

$$\alpha^{\frac{P-1}{Q-1}}.$$

Now Theorem 10 together with the supplement can be reworded without difficulty as follows:

THEOREM 11. (Explicit formulae for the regular solutions of Problem III.) *Suppose that a pair* $P, \varphi(x)$ *with*

$$P \underset{1}{|} q, \quad \varphi(x) \underset{1}{\in} \mathsf{F}[x], \quad \varphi(x)|x^P - x, \quad \frac{P-1}{2} \leqq \varphi^\circ \leqq P-1$$

*is given. Let us take numbers* $P_0, \ldots, P_k$ *with*

$$P_0 \underset{1}{|} P_1 \underset{1}{|} \cdots \underset{1}{|} P_k \underset{1}{|} q, \qquad P = P_0 < P_1 < \cdots < P_k = q,$$

*then (not necessarily different) zeros*

$$\alpha_0, \ldots, \alpha_{k-1} \in \mathsf{F}_P$$

*of the cofactor* $\bar{\varphi}(x)\big(=\varphi(x)^{-1}(x^P - x)\big)$ *and finally in general not uniquely determined elements*

$$\varrho_0 = \sigma_0, \quad \varrho_i = \Big(\Big(\ldots\big((\alpha_i - \varrho_0)^{\frac{P_0-1}{P_1-1}} - \varrho_1\big)\ldots\Big)^{\frac{P_{i-2}-1}{P_{i-1}-1}} - \varrho_{i-1}\Big)^{\frac{P_{i-1}-1}{P_i-1}} \; (\in \mathsf{F}_{P_i})$$
$$(i = 1, \ldots, k-1)$$

*and an arbitrary* $\varrho_R \in \mathsf{F}_{P_k} (= \mathsf{F})$. *Then the polynomials*

$$f(x) = \Big(\ldots\big((x + \varrho_k)^{\frac{P_k-1}{P_{k-1}-1}} + \varrho_{k-1}\big)\ldots\Big)^{\frac{P_1-1}{P_0-1}} + \varrho_0,$$

*formed from these elements, are all the regular solutions of Problem III, belonging to* $P, \varphi(x)$. *This holds also under the restriction*

$$\varrho_1, \ldots, \varrho_{k-1} \neq 0,$$

*which yields a unique representation of all regular solutions.*

NOTE. Clearly $k$ in this theorem can be any non negative integer not exceeding the number of prime factors of $n$, the degree of $\mathsf{F}$.

EXAMPLE. For $k=3$ we have (*inter alia*)

$$\varrho_0 = \alpha_0,$$

$$\varrho_1 = (\alpha_1 - \varrho_0)^{\overline{\frac{P_0-1}{P_1-1}}},$$

$$\varrho_2 = \left((\alpha_2 - \varrho_0)^{\overline{\frac{P_0-1}{P_1-1}}} - \varrho_1\right)^{\overline{\frac{P_1-1}{P_2-1}}}$$

and

$$f(x) = \left(\left(\left(x+\varrho_0\right)^{\overline{\frac{P_3-1}{P_2-1}}} + \varrho_2\right)^{\overline{\frac{P_2-1}{P_1-1}}} + \varrho_1\right)^{\overline{\frac{P_1-1}{P_0-1}}} + \varrho_0$$

## § 27. Another characterization of the non-primitive and primitive solutions of Problem III

We shall make considerable progress in Problem III if we characterize its non-primitive solutions—as has been promised—in an alternative way (whence a new characterization of the primitive solutions will also come about automatically).

For our purposes let us assume a non-linear solution $f(x)$ of Problem III to be given belonging to the pair $P$, $\varphi(x)$.

We first assume $f(x)$ to be non-primitive. An easily deducible necessary condition for this will prove later to be sufficient too.

Our assumption implies

$$f(x) = g(x)^{\overline{\frac{Q-1}{P-1}}} + \varrho \tag{1}$$

with

$$f(x), g(x) \underset{1}{\in} \mathsf{F}[x]; \quad f^\circ = \frac{q-1}{P-1}, \quad g^\circ = \frac{q-1}{Q-1}; \quad f^{\circ_P} = g^{\circ_Q} = 1 \tag{2}$$

$$P \underset{1}{|} Q \underset{1}{|} q; \quad P < Q (\leqq q) \tag{3}$$

and

$$\varrho \in \mathsf{F}_P, \quad \varphi(\varrho) \neq 0. \tag{4}$$

(Although $\varphi(f(x)) | x^q - x$ is also true, this will not be used here.)

We next calculate the two-step constituents $f_{ij}(x)$ $(i=0, 1)$ of $f(x)$. By Proposition 25, the latter are characterized by

$$f(x) = f_{00}(x) + x^{q_2} f_{01}(x) + x^{q_1} f_{10}(x) + x^{q_1+q_2} f_{11}(x) \left(f_{ij}(x) \in \mathsf{F}[x];\ i, j = 0, 1\right) \tag{5}$$

and

$$f_{ij}^\circ < q_2 \qquad (i, j = 0, 1). \tag{6}$$

On the other hand

$$g(x) = g_0(x)^Q + x^{\overline{\frac{q}{Q}}} g_1(x)^Q \tag{7}$$

holds (by (2)) with

$$g_0(x), g_1(x) \in \mathsf{F}[x]; \quad g_0^\circ \leqq g_1^\circ = \frac{Q^{-1}q-1}{Q-1}. \tag{8}$$

After we have put

$$G(x) = g(x)^{\frac{P^{-2}Q-1}{P-1}} \tag{9}$$

(1) can be written in the form

$$f(x) = g(x)^{\frac{Q}{P}} g(x)^{\frac{Q}{P^2}} G(x) + \varrho,$$

consequently we obtain from (7) that

$$f(x) = \left(g_0(x)^{\frac{Q}{P}} + x^{\frac{q}{P}} g_1(x)^{\frac{Q}{P}}\right)\left(g_0(x)^{\frac{Q}{P^2}} + x^{\frac{q}{P^2}} g_1(x)^{\frac{Q}{P^2}}\right) G(x) + \varrho.$$

Considering $\frac{q}{P} = q_1$, $\frac{q}{P^2} = q_2$ and multiplying out, comparison with (5) leads to the four equations

$$\begin{aligned} f_{00}(x) &= g_0(x)^{\frac{Q}{P}+\frac{Q}{P^2}} G(x) + \varrho, & f_{01}(x) &= g_0(x)^{\frac{Q}{P}} g_1(x)^{\frac{Q}{P^2}} G(x),\\ f_{10}(x) &= g_0(x)^{\frac{Q}{P^2}} g_1(x)^{\frac{Q}{P}} G(x), & f_{11}(x) &= g_1(x)^{\frac{Q}{P}+\frac{Q}{P^2}} G(x), \end{aligned} \tag{10}$$

but it still remains to be proved that the degree estimations (6) are valid for (10). Indeed, (7), ($8_2$), (9) and (10) imply

$$f_{ij}^\circ \leqq \left(\frac{Q}{P} + \frac{Q}{P^2}\right) \frac{Q^{-1}q-1}{Q-1} + \frac{P^{-2}Q-1}{P-1} \frac{q-1}{Q-1} = \frac{P^{-2}q-1}{P-1} = \frac{q_2-1}{P-1},$$

whence (6) follows.

We use (10) for the calculation of the determinant

$$D(x) = \begin{vmatrix} f_{00}(x) & f_{01}(x) \\ f_{10}(x) & f_{11}(x) \end{vmatrix}$$

(of $f(x)$). Clearly $D(x) = \varrho f(x)$, i.e.

$$D(x) - \varrho f_{11}(x) = 0.$$

However, the left-hand side is, as a consequence of (4), just the codeterminant $D_\varrho(x)$ of $f(x)$ (belonging to $\varrho$). This proves the "only if" part of our promised

THEOREM 12. (Criterion for the non-primitive solutions of Problem III.) *A non-linear solution $f(x)$ of Problem III is non-primitive if and only if one of the codeterminants $D_\varrho(x)$ of $f(x)$ vanishes.*

COROLLARY. *$f(x)$ is primitive if and only if its codeterminants are different from* 0.

The corollary follows from the theorem. We have to prove the "if" part of the theorem, but the proof will be rather tricky and it will require not only §§ 21, 22, 23 but also some other preliminary results.

As far as the solution $f(x)$, belonging to $P$, $\varphi(x)$ and considered at the outset, is concerned, we now suppose that one of the codeterminants $D_\varrho(x)$ vanishes. This means the validity of

$$D(x) - \varrho f_{11}(x) = 0 \tag{11}$$

for suitable $\varrho$ with (4). We have to prove that $f(x)$ is non-primitive.

We first discuss the case

$$\varrho = 0 \tag{12}$$

to which the general case will easily reduce. In view of (11), assumption (12) asserts that

$$D(x) = 0. \tag{13}$$

An apparently insignificant but essential simplification will be achieved by subjecting $f(x)$ to a translation

$$x \to x + \xi \qquad (\xi \in \mathsf{F}).$$

From Propositions 26 and 29 this is certainly allowed, since according to these the determinant of $f(x+\xi)$, on the one hand, is equal to $D(x+\xi)$ and thus, thanks to (13), also vanishes. On the other hand $f(x+\xi)$, too, is non-primitive. (Here $f(x+\xi)$ still belongs unchanged to the pair $P$, $\varphi(x)$.) This implies, by virtue of Proposition 2, that we may also assume that

$$f^{\downarrow} \geqq 2. \tag{14}$$

(Without this simplifying assumption we should be unable to carry out our proof.)

We shall use the relative degree

$$m = (\mathsf{F} | \mathsf{F}_p)^\circ \, (\geqq 2). \tag{15}$$

It follows that

$$q = P^m. \tag{16}$$

We put (as a generalization of the notations $q_1$, $q_2$, introduced in § 23)

$$q_i = \frac{q}{P^i} = P^{m-i} \qquad (i = 1, \ldots, m;\ q > q_1 > \cdots > q_m = 1). \tag{17}$$

We introduce the set

$$\mathscr{E} = \{0, 1, P, P+1, P^2, P^2+1, P^2+P, P^2+P+1, P^3, \ldots\}$$

of non-negative integers of the $P$-adic height $\leqq 1$. Note that a polynomial from $\mathsf{F}[x]$ is $P$-linear if and only if its exponent set lies in $\mathscr{E}$. Furthermore, let

$$\mathscr{E}_s = \mathscr{E} \cap \{0, 1, \ldots, P^s-1\} = \{i \in \mathscr{E}\colon\ i < P^s\} \qquad (s = 0, \ldots, m;\ \mathscr{E}_0 = 0).$$

Then

$$\max \mathscr{E}_s = \frac{P^s-1}{P-1} = P^{s-1}+\cdots+P+1 = q_{m-s+1}+q_{m-s+2}+\cdots+q_m$$
$$(s = 0, \ldots, m).$$

In particular,

$$\mathscr{E}_m = \mathscr{E} \cap \{0, \ldots, q-1\} = \{i \in \mathscr{E};\ i < q)$$

holds with

$$\max \mathscr{E}_m = \frac{q-1}{P-1} = q_1+\cdots+q_m.$$

It should be noted that the partial sums of the right-hand side are in fact all the elements of $\mathscr{E}_m$.

We introduce a *semiordering* $\preceq$ into $\mathscr{E}$, defined for arbitrary $i, j \in \mathscr{E}$ by

$$i \preceq j \Leftrightarrow j-i \in \mathscr{E}.$$

Let $i \prec j$ signify the same as $i \preceq j$, $i \neq j$. In this case $i$ will be said to lie *below* $j$ (and $j$ *above* $i$). It is clear that, $j \in \mathscr{E}$ being fixed, all the $i \prec j$ are given by cancelling some (at least one) of the $P$-adic numerals of $j$ (i.e. replacing them by 0). We see that $\mathscr{E}$ is a lattice and every $\mathscr{E}_s$ is a sublattice of it. (It should be noted that $\mathscr{E}$ and $\mathscr{E}_s$ are isomorphic to the lattices of subsets of $\mathscr{N}$ and $\{0. \ldots. s-1\}$, respectively.)

By our assumption we have that

$$f(x) \underset{1}{\in} \mathsf{F}[x], \quad f^\circ = \frac{q-1}{P-1}, \quad f^{\circ_P} = 1, \quad x^q - x \mid f(x)^P - f(x). \tag{18}$$

Accordingly we write the coefficient representation of $f(x)$ in the form

$$f(x) = \sum_{l \in \mathscr{E}_m} \alpha_l x^l \qquad \left(\alpha_l \in \mathsf{F};\ \alpha_{\frac{q-1}{P-1}} = 1\right). \tag{19}$$

Here, from $(18_4)$, Theorem 2 and Proposition 6 the conjugacy conditions

$$\alpha_{l\sqcap} = \alpha_l^P \qquad (l \in \mathscr{E}_m) \tag{20}$$

hold. It should be noted that although the cyclic permutation $\sqcap$ (of $P$-adic numerals) has been defined for the elements of $\{0, \ldots, q-1\}$, in what follows

$\sqcap$ will be applied (as already in (20)) only to the elements of the subset $\mathscr{E}_m$. It should be emphasized that this does not take us out of $\mathscr{E}_m$, i.e. $l \in \mathscr{E}_m$ always implies $l^\sqcap \in \mathscr{E}_m$. (For the $\mathscr{E}_s$ with $0<s<m$ the corresponding statement does not, of course, hold.)

By virtue of (14)

$$\alpha_{\frac{q-1}{P-1}-1}=0$$

holds for the coefficients in (19). Hence and from (20), more generally, it follows that

$$\alpha_{\frac{q-1}{P-1}-q_i}=0. \tag{21}$$

A sum $A+B$ of complexes $A$ and $B$ of a module is said to be *schlicht* and denoted by $A\dot{+}B$ if all its elements can be written uniquely in the form

$$a+b \qquad (a\in A,\ b\in B).$$

(This is clearly an immediate generalization of the direct sum of submodules. Its multiplicative analogue, viz. the "schlicht product" of complexes of a finite abelian group is a central concept of Hajós's theory and it will occur in Chapter VI.)

Obviously

$$\mathscr{E}_m=\{q_1, q_2\}\dot{+}\mathscr{E}_{m-2}$$

and therefore (19) can be decomposed into

$$f(x) = \sum_l \alpha_l x^l + \sum_l \alpha_{q_1+l} x^{q_1+l} + \sum_l \alpha_{q_2+l} x^{q_2+l} + \sum_l \alpha_{q_1+q_2+l} x^{q_1+q_2+l} \qquad (l\in\mathscr{E}_{m-2}),$$

or, more concisely,

$$f(x) = \sum_{i,j=0}^{1} \Big(x^{iq_1+jq_2} \sum_{l\in\mathscr{E}_{m-2}} \alpha_{iq_1+jq_2+l} x^l\Big).$$

The inner sum is of formal degree

$$\max \mathscr{E}_{m-2} = \frac{P^{m-2}-1}{P-1} = \frac{q_2-1}{P-1},$$

consequently comparison with (5) and (6) leads to

$$f_{ij}(x) = \sum_{l\in\mathscr{E}_{m-2}} \alpha_{iq_1+jq_2+l} x^l \qquad (i,j=0,1). \tag{22}$$

Before making use of (22) in (13), we go from $f(x)$ over to the reciprocal polynomial (cf. ($18_1$) and ($18_2$))

$$\bar{f}(x)=x^{\frac{q-1}{P-1}} f\left(\frac{1}{x}\right) \qquad (f(0)=1), \tag{23}$$

which leads to considerable formal simplification in what follows. At the same time we introduce the polynomials

$$\bar{f}_{ij}(x) = x^{\frac{q_2-1}{P-1}} f_{1-i,1-j}\left(\frac{1}{x}\right) \qquad (i, j = 0, 1;\ \bar{f}_{11}(0) = 1), \tag{24}$$

reciprocal to (22). (Here the indices are transformed by $i \to 1-i$ and $j \to 1-j$, the point of which will soon be clear.)

From (19) and (23) we have:

$$\bar{f}(x) = \sum_{l \in \mathscr{E}_m} \alpha_l x^{\frac{q-1}{P-1}-l}.$$

Since $l \to \dfrac{q-1}{P-1} - l$ is a permutation of the set $\mathscr{E}_m$ which may reasonably be called its *reflection*, it follows that

$$\bar{f}(x) = \sum_{l \in \mathscr{E}_m} \beta_l x^l \tag{25}$$

holds with

$$\beta_l = \alpha_{\frac{q-1}{P-1}-l} \qquad (l \in \mathscr{E}_m). \tag{26}$$

By (19), (20), (21) we also have

$$\beta_0 = 1, \quad \beta_{l\sqcap} = \beta_l^P \quad (l \in \mathscr{E}_m), \quad \beta_{q_i} = 0 \qquad (i = 1, \ldots, m). \tag{27}$$

From (22), (24) we infer that

$$\bar{f}_{ij}(x) = \sum_{l \in \mathscr{E}_{m-2}} \alpha_{(1-i)q_1 + (1-j)q_2 + l}\, x^{\frac{q_2-1}{P-1}-l} \qquad (i, j = 0, 1).$$

Let us execute the reflection

$$l \to \frac{q_2-1}{P-1} - l$$

of $\mathscr{E}_{m-2}$, applying (26) and

$$q_1 + q_2 + \frac{q_2-1}{P-1} = \frac{q-1}{P-1}.$$

Thus we obtain

$$f_{ij}(x) = \sum_{l \in \mathscr{E}_{m-2}} \beta_{iq_1 + jq_2 + l}\, x^l \qquad (i, j = 0, 1) \tag{28}$$

(similarly to (22)). This shows that the $f_{ij}(x)$ are $P$-linear, a result which may also be obtained directly from (24).

When we apply (24) the determinant

$$D(x) = \begin{vmatrix} f_{00}(x) & f_{01}(x) \\ f_{10}(x) & f_{11}(x) \end{vmatrix}$$

after the transformation $x \to \dfrac{1}{x}$, followed by reflection in the diagonal and the secondary diagonal and, finally, after multiplication by a suitable power of $x$ becomes the determinant $|\bar{f}_{ij}(x)|$. Consequently, by (13)

$$\begin{vmatrix} \bar{f}_{00}(x) & \bar{f}_{01}(x) \\ \bar{f}_{10}(x) & \bar{f}_{11}(x) \end{vmatrix} = 0.$$

Hence and from Lemma 2 the existence of four $P$-linear polynomials

$$a(x), \ldots, d(x) \in \mathsf{F}[x]$$

follows such that

$$\begin{aligned} \bar{f}_{00}(x) &= a(x) \times b(x), \quad \bar{f}_{11}(x) = c(x) \times d(x), \\ \bar{f}_{01}(x) &= a(x) \times c(x), \quad \bar{f}_{10}(x) = b(x) \times d(x). \end{aligned} \tag{29}$$

As (cf. (24)) $f_{00}(0) = 1$, in view of $(29_1)$ we are allowed to assume

$$a(0) = b(0) = 1. \tag{30}$$

(25), (28) clearly imply that

$$\bar{f}(x) = \bar{f}_{00}(x) + x^{q_2}\bar{f}_{01}(x) + x^{q_1}\bar{f}_{10}(x) + x^{q_1+q_2}\bar{f}_{11}(x).$$

In view of the degree estimations

$$\bar{f}_{ij}^{\circ} < q_2 \qquad (i, j = 0, 1)$$

(valid by (5), (6) and (23)) substitution of (29) leads to the direct decomposition

$$\bar{f}(x) = \big(a(x) + x^{q_1} d(x)\big) \times \big(b(x) + x^{q_2} c(x)\big). \tag{31}$$

It is clear that here both factors are again $P$-linear.

It is obvious that a product $A(x)B(x)$ of polynomials is direct if and only if the schlicht sum $\mathfrak{E}(A) \dotplus \mathfrak{E}(B)$ of the exponent sums of the factors exists. Therefore (25) and (31) imply the existence of a decomposition

$$\frac{q-1}{P-1} (= q_1 + \cdots + q_m) = u + v \qquad (u, v \in \mathscr{E}_m) \tag{32}$$

such that

$$q_1 \leqq u, \quad q_2 \leqq v \tag{33}$$

holds and (31) can be written with suitable $\kappa_i$, $\lambda_j$ in the form

$$\sum_{l \in \mathscr{E}_m} \beta_l x^l = \Big(\sum_{i \leqq u} \kappa_i x^i\Big)\Big(\sum_{j \leqq v} \lambda_j x^j\Big) \tag{34}$$

where, by (30), we may also assume that

$$\kappa_0 = \lambda_0 = 1. \tag{35}$$

(The $a(x, \ldots, d(x)$ will no more be used.)

As, however, from (32) the elements $l \in \mathscr{E}_m$ (i.e. the $l \leqq q_1 + \cdots + q_m$) are uniquely represented by $i+j$ $(i \leqq u,\ j \leqq v)$, after multiplying out in (34) the validity of all the relations

$$\beta_{i+j} = \kappa_i \lambda_j \qquad (i \leqq u,\ j \leqq v)$$

follows. In particular, for $j=0$ and $i=0$ we simply obtain from this the validity of $\beta_i = \kappa_i$ $(i \leqq u)$ and $\beta_j = \lambda_j$ $(j \leqq v)$, respectively. Thus we finally have the relations

$$\beta_{i+j} = \beta_i \beta_j \qquad (i \leqq u,\ j \leqq v). \tag{36}$$

In order to draw further conclusions from this (and from (27)), we change the notation for the $\beta_l$. To this end we put

$$\mathfrak{M} = (\mathscr{Z}/m)^+, \tag{37}$$

where $\mathscr{Z}/m$ itself is understood as the residue class ring of $\mathscr{Z}$ mod $m$, and so $\mathfrak{M}$ denotes the module (of the elements) of this ring.

For convenience we shall interpret the indices $1, \ldots, m$, that appear in the $q_1, \ldots, q_m$, also as elements of $\mathfrak{M}$. However, to avoid any misunderstanding, it is advisable to denote the elements of $\mathfrak{M}$ (consequently also those of $\mathscr{Z}/m$) by $\varepsilon, 2\varepsilon, \ldots, m\varepsilon$, so that $\varepsilon$ is the unity element (i.e. the residue class 1 (mod $m$)). Our convention implies the validity of

$$q_k = q_{k\varepsilon} \qquad (k = 1, \ldots, m). \tag{38}$$

(On the right-hand side $k$ is considered, of course, only mod $m$.)

It follows that the $l \in \mathscr{E}_m$ can be uniquely represented by the sums

$$\sum_{\mu \in \mathfrak{C}} q_\mu \qquad (\mathfrak{C} \subseteqq \mathfrak{M}). \tag{39}$$

(As indicated, here $\mathfrak{C}$ runs through all complexes of $\mathfrak{M}$, including the empty complex $\varnothing$.)

We now introduce the promised change of notation for the coefficients $\beta_l$ (corresponding to (39)) by putting

$$\beta(\mathfrak{C}) = \beta_l \qquad (\mathfrak{C} \subseteqq \mathfrak{M},\ l = \sum_{\mu \in \mathfrak{C}} q_\mu). \tag{40}$$

(The usefulness of this notation will soon be clear.)

We consider an arbitrary $l \in \mathscr{E}_m$, writing it, as in (40), in the form

$$l = \sum_{\mu \in \mathfrak{C}} q_\mu .$$

Clearly (cf. (17))

$$l^{\sqcap} = \sum_{\mu \in \mathfrak{C}} q_{\mu-\varepsilon}$$

follows. Because of this and of (40), ($27_2$) becomes

$$\beta(-\varepsilon+\mathfrak{C}) = \beta(\mathfrak{C})^P.$$

(Of course, $-\varepsilon+\mathfrak{C}$ is to be interpreted as the sum of the complexes $-\varepsilon$ and $\mathfrak{C}$; it consists therefore of all the $-\varepsilon+\mu$ with $\mu\in\mathfrak{C}$.) We have thus obtained the more general rule

$$\beta(-k\varepsilon+\mathfrak{C}) = \beta(\mathfrak{C})^{Pk} \qquad (k = 0, 1, \ldots; \mathfrak{C}\subseteqq\mathfrak{M}). \tag{41}$$

We note that this implies the equivalence

$$\beta(\mathfrak{C}) = 0 \Leftrightarrow \beta(\mu+\mathfrak{C}) = 0 \tag{42}$$

for all $\mu\in\mathfrak{M}$ and $\mathfrak{C}\subseteqq\mathfrak{M}$.

By ($27_1$) and ($27_3$)

$$\beta(\varnothing)=1, \quad \beta(\mu)=0 \qquad (\mu\in\mathfrak{M}). \tag{43}$$

If

$$f(x)=x^{\frac{q-1}{P-1}}$$

$f(x)$ is (because $P<q$) obviously non-primitive; thus we shall henceforth be able to exclude this case. (It will be seen, *inter alia*, that in the remaining case $m$ must be a composite number.) Since $\bar{f}(x)$ now has at least two terms, it follows from (23) that the same holds for $f(x)$. There is therefore at least one coefficient $\beta(\mathfrak{C})$ with $\mathfrak{C}\neq\varnothing$ different from 0. We denote with $\mathfrak{N}$ a non-empty complex of $\mathfrak{M}$ with minimal cardinality $O(\mathfrak{N})$ for which

$$\beta(\mathfrak{N})\neq 0 \tag{44}$$

holds. As a consequence of ($27_3$),

$$O(\mathfrak{N})\geqq 2 \tag{45}$$

must hold. As by (42) and (44) all the

$$\beta(\mu+\mathfrak{N})\neq 0 \qquad (\mu\in\mathfrak{M}) \tag{46}$$

are valid, we may assume that

$$0\in\mathfrak{N}. \tag{47}$$

We can write the above $u$, $v$ (cf. (32)) in the form

$$u=\sum_{\mu\in\mathfrak{U}} q_\mu, \quad v=\sum_{\mu\in\mathfrak{B}} q_\mu,$$

where $\mathfrak{U}$, $\mathfrak{V}$ are suitable complementary complexes of $\mathfrak{M}$, i.e. those with

$$\mathfrak{U}\cup\mathfrak{V}=\mathfrak{M}, \quad \mathfrak{U}\cap\mathfrak{V}\neq\varnothing \tag{48}$$

and, by (33) (cf. also (38))

$$\varepsilon\in\mathfrak{U}, \quad 2\varepsilon\in\mathfrak{V} \tag{49}$$

holds.

Then by (36) and (40) we have

$$\beta(\mathfrak{A}\cup\mathfrak{B})=\beta(\mathfrak{A})\beta(\mathfrak{B}) \qquad (\mathfrak{A}\subseteqq\mathfrak{U},\ \mathfrak{B}\subseteqq\mathfrak{V}).$$

By virtue of (48), this can immediately be replaced by

$$\beta(\mathfrak{C})=\beta(\mathfrak{U}\cap\mathfrak{C})\beta(\mathfrak{V}\cap\mathfrak{C}) \qquad (\mathfrak{C}\subseteqq\mathfrak{M}). \tag{50}$$

Applying (50) with $\mathfrak{C} = \mu+\mathfrak{N}$, we obtain

$$\beta(\mathfrak{U}\cap(\mu+\mathfrak{N}))\beta(\mathfrak{V}\cap(\mu+\mathfrak{N})) \neq 0 \qquad (\mu\in\mathfrak{M}) \tag{51}$$

using (46).

On the other hand, by (48) (and $O(\mu+\mathfrak{N}) = O(\mathfrak{N})$),

$$O(\mathfrak{U}\cap(\mu+\mathfrak{N}))+O(\mathfrak{V}\cap(\mu+\mathfrak{N})) = O(\mathfrak{N}).$$

As a consequence of (51) and the minimum property of $\mathfrak{N}$, a term must vanish on the left-hand side (the other one must be equal to $O(\mathfrak{N})$). Hence we infer that, for every $\mu$, $\mu+\mathfrak{N}$ must be contained either in $\mathfrak{U}$ or in $\mathfrak{V}$. Thus for the elements $\mu\in\mathfrak{M}$ the implication

$$\mathfrak{U}\cap(\mu+\mathfrak{N}) \neq \varnothing \Rightarrow \mu+\mathfrak{N} \subseteqq \mathfrak{U}$$

holds. After having carried out the substitution $\mu \rightarrow -\mu$ and a few simple transformations we obtain

$$(\mu+\mathfrak{U})\cap\mathfrak{N} \neq \varnothing \Rightarrow \mu+\mathfrak{U} \supseteqq \mathfrak{N}.$$

According to this result and to (47), the additive form of Proposition 18 can be applied to $\mathfrak{M}$, $\mathfrak{U}$, $\mathfrak{N}$ (in place of $G$, $K$, $L$), leading to the fact that $\mathfrak{U}$ is the union of classes mod $(\mathfrak{N})$, where $(\mathfrak{N})$ denotes the submodule of $\mathfrak{M}$, generated by (the elements of) $\mathfrak{N}$. It should be noted that it is certainly true that $(\mathfrak{N})\neq 0$, from (45). It follows from these two statements that the complex $\mathfrak{U}$ is non-primitive. (This result was obtained with the aid of $\mathfrak{N}$; in what follows, however, $\mathfrak{N}$ will no longer be used.)

We denote by $\mathfrak{W}$ the maximal submodule of $\mathfrak{M}$ with the property that $\mathfrak{U}$ (and thus, by (48), also $\mathfrak{V}$) is the union of classes mod $\mathfrak{W}$. According to the result just obtained we have $\mathfrak{W}\neq 0$. Also $\mathfrak{W}\neq\mathfrak{M}$, this being the consequence of the fact that ($49_2$) implies $\mathfrak{B} \neq \emptyset$. Thus $\mathfrak{W}$ is a proper submodule of $\mathfrak{M}$. This means (cf. (37)) that

$$1 < O(\mathfrak{W}) < m (= O(\mathfrak{M})). \tag{52}$$

(It follows from this, of course, that $m$ is composite, which was promised above.)

The order and the index of $\mathfrak{W}$ will be written as $e$ and $e'$, respectively:

$$e = O(\mathfrak{W}), \quad e' = O(\mathfrak{M}/\mathfrak{W}) \qquad (ee' = m;\ e, e' > 1). \tag{53}$$

Hence

$$\mathfrak{W} = \{0, e'\varepsilon, 2e'\varepsilon, \ldots, (e-1)e'\varepsilon\}; \tag{54}$$

moreover, it follows that

$$\mathfrak{R} = \{0, \varepsilon, 2\varepsilon, \ldots, (e'-1)\varepsilon\} \tag{55}$$

is a representative system of $\mathfrak{M}$ mod $\mathfrak{W}$. This implies the validity of the schlicht sum decomposition

$$\mathfrak{M} = \mathfrak{R} \dot{+} \mathfrak{W}. \tag{56}$$

Let us decompose $\mathfrak{R}$ into two disjoint complexes $\mathfrak{S}$ and $\mathfrak{T}$ in such a way that each of the latter should form a representative system of the classes mod $\mathfrak{W}$ of which $\mathfrak{U}$ and $\mathfrak{B}$ consist, respectively. This is possible, by virtue of the definition of $\mathfrak{W}$ and $\mathfrak{R}$, and from (48). Thus we have

$$\mathfrak{S} \cup \mathfrak{T} = \mathfrak{R}, \quad \mathfrak{S} \cap \mathfrak{T} = \emptyset \tag{57}$$

$$\mathfrak{U} = \mathfrak{S} + \mathfrak{W}, \quad \mathfrak{B} = \mathfrak{T} + \mathfrak{W}. \tag{58}$$

By $\mathfrak{U}/\mathfrak{W}$ we here denote the set of the classes mod $\mathfrak{W}$ of which $\mathfrak{U}$ consists and consider $\mathfrak{U}/\mathfrak{W}$ as a complex of the factor module $\mathfrak{M}/\mathfrak{W}$. Because of the maximal property of $\mathfrak{W}$ the factor module $\mathfrak{N}/\mathfrak{W}$ is a primitive complex, consequently the additive form of Proposition 19 can be applied to $\mathfrak{M}/\mathfrak{W}$, $\mathfrak{U}/\mathfrak{W}$ (in place of $G$, $K$). Thus by ($58_1$) we get

$$\bigcap_{\sigma \in \mathfrak{S}} (-\sigma + \mathfrak{U}) = \mathfrak{W}. \tag{59}$$

(The same holds with $\mathfrak{T}$, $\mathfrak{B}$ instead of $\mathfrak{S}$, $\mathfrak{U}$, a fact which we shall not, however, use.)

Let us now consider an arbitrary coefficient

$$\beta(\mathfrak{C}) \qquad (\mathfrak{C} \subseteqq \mathfrak{M}) \tag{60}$$

of $\bar{f}(x)$ (cf. (25) and (40)). We put

$$\mathfrak{C}_\sigma = (\sigma + \mathfrak{W}) \cap \mathfrak{C} \qquad (\sigma \in \mathfrak{R}) \tag{61}$$

and prove the "decomposition formula"

$$\beta(\mathfrak{C}) = \prod_{\sigma \in \mathfrak{R}} \beta(\mathfrak{C}_\sigma). \tag{62}$$

We remark, first of all, that the complexes $\mathfrak{C}_\sigma$ are, by virtue of (56) and (61), disjoint by pairs and that

$$\mathfrak{C} = \bigcup_{\sigma \in \mathfrak{R}} \mathfrak{C}_\sigma \tag{63}$$

holds; for this reason we shall call them the *components* of $\mathfrak{C}$, for short.

Let $k$ denote the number of the components $\mathfrak{C}_\sigma$, different from $\varnothing$. If $k=0$, i.e. if all the $\mathfrak{C}_\sigma$ are equal to $\varnothing$, then on account of (63) $\mathfrak{C}=\varnothing$, and so (62) now holds because of ($43_1$). When $k=1$ $\mathfrak{C}$ is equal to a $\mathfrak{C}_\sigma(\neq\varnothing)$, and therefore (62) is again correct, the proof being slightly different. In the remaining case $k \geqq 2$ we suppose the assertion to be true for the lesser values of $k$.

It follows from (58) and definition (61) that the intersections $\mathfrak{U} \cap \mathfrak{C}$ and $\mathfrak{B} \cap \mathfrak{C}$ are unions of certain components of $\mathfrak{C}$. Also

$$(\mathfrak{U} \cap \mathfrak{C}) \cup (\mathfrak{B} \cap \mathfrak{C}) = \mathfrak{C}, \quad (\mathfrak{U} \cap \mathfrak{C}) \cap (\mathfrak{B} \cap \mathfrak{C}) = \varnothing$$

is a consequence of (48). Thus, if $i$ and $j$ denote the number of components of $\mathfrak{C}$ different from $\varnothing$ and appearing in $\mathfrak{U} \cap \mathfrak{C}$ or in $\mathfrak{B} \cap \mathfrak{C}$, respectively, it follows that

$$i+j = k.$$

When $i, j<k$ (62) follows by induction for $\mathfrak{U} \cap \mathfrak{C}$ and $\mathfrak{B} \cap \mathfrak{C}$ (in place of $\mathfrak{C}$). Therefore, if the factors of the right-hand side of (50) are replaced by their product decompositions as in (62), formula (62) emerges, which was to be proved.

In the remaining case either $i$ or $j$ is equal to $k$ (the other equal to 0). By symmetry it is sufficient to consider the case $i=k$, $j=0$. Then $\mathfrak{U} \cap \mathfrak{C} = \mathfrak{C}$ and thus

$$\mathfrak{C} \subseteqq \mathfrak{U}. \tag{64}$$

Let $\mathfrak{C}_{\sigma_1}$, $\mathfrak{C}_{\sigma_2}$ be two different components $\neq\varnothing$ of $\mathfrak{C}$ ($\sigma_1, \sigma_2 \in \mathfrak{R}$; $\sigma_1 \neq \sigma_2$). By (61) we have

$$\mathfrak{C}_{\sigma_s} \subseteqq \sigma_s + \mathfrak{W} \qquad (s = 1, 2). \tag{65}$$

From (59) and $\sigma_2 - \sigma_1 \notin \mathfrak{W}$ (also using the fact that $i \geqq 2$) we infer the existence of a $\bar{\sigma} \in \mathfrak{S}$ with

$$\mathfrak{W} \subseteqq -\bar{\sigma} + \mathfrak{U} \qquad (\sigma_2 - \sigma_1 + \mathfrak{W}) \cap (-\bar{\sigma} + \mathfrak{U}) \neq \varnothing. \tag{66}$$

According to (50),

$$\beta(\bar{\sigma} - \sigma_1 + \mathfrak{C}) = \beta(\mathfrak{D}_1)\beta(\mathfrak{D}_2) \tag{67}$$

holds with

$$\mathfrak{D}_1 = \mathfrak{U} \cap (\bar{\sigma} - \sigma_1 + \mathfrak{C}), \quad \mathfrak{D}_2 = \mathfrak{V} \cap (\bar{\sigma} - \sigma_1 + \mathfrak{C}) \tag{68}$$

whence from (48)

$$\bar{\sigma} - \sigma_1 + \mathfrak{C} = \mathfrak{D}_1 \cup \mathfrak{D}_2, \quad \mathfrak{D}_1 \cap \mathfrak{D}_2 = \varnothing \tag{69}$$

follows.

By ($68_1$)

$$-\bar{\sigma} + \mathfrak{D}_1 = (-\bar{\sigma} + \mathfrak{U}) \cap (-\sigma_1 + \mathfrak{C}). \tag{70}$$

Now, $\mathfrak{C}_{\sigma_1} \subset \mathfrak{C}$, so that

$$-\sigma_1 + \mathfrak{C}_{\sigma_1} \subset \sigma_1 + \mathfrak{C}.$$

On the other hand, from the case $s=1$ of (65) and from ($66_1$) we deduce that

$$-\sigma_1 + \mathfrak{C}_{\sigma_1} (\subseteqq \mathfrak{W}) \subseteqq -\bar{\sigma} + \mathfrak{U}.$$

Thus from (70) we obtain $-\sigma_1 + \mathfrak{C}_{\sigma_1} \subseteqq -\bar{\sigma} + \mathfrak{D}_1$, i.e.

$$\bar{\sigma} - \sigma_1 + \mathfrak{C}_{\sigma_1} \subseteqq \mathfrak{D}_1. \tag{71}$$

From the case $s=2$ of (65) it follows that

$$-\sigma_1 + \mathfrak{C}_{\sigma_2} \subseteqq \sigma_2 - \sigma_1 + \mathfrak{W},$$

moreover, by (70) we have

$$-\bar{\sigma} + \mathfrak{D}_1 \subseteqq -\bar{\sigma} + \mathfrak{U}$$

and so by ($66_2$)

$$(-\sigma_1 + \mathfrak{C}_{\sigma_2}) \cap (-\bar{\sigma} + \mathfrak{D}_1) = \varnothing,$$

i.e.

$$(\bar{\sigma} - \sigma_1 + \mathfrak{C}_{\sigma_2}) \cap \mathfrak{D}_1 = \varnothing. \tag{72}$$

According to definition (61), all the components of

$$\bar{\sigma} - \sigma_1 + \mathfrak{C} \tag{73}$$

are clearly

$$\bar{\sigma} - \sigma_1 + \mathfrak{C}_\sigma \qquad (\sigma \in \mathfrak{R}). \tag{74}$$

Furthermore, since $\mathfrak{U}$ is the union of classes mod $\mathfrak{W}$, it follows from (68) and (69) that the components of $\mathfrak{D}_1$ and $\mathfrak{D}_2$ (taken together) are exactly those of (73), i.e. all the complexes (74). Since, however, (74) is different from $\varnothing$ for $\sigma = \sigma_1$ ($s = 1, 2$), it follows from (71) and (72) that the number of components of $\mathfrak{D}_1$, different from $\varnothing$, is at least 1 and less than $k$. The same also holds for $\mathfrak{D}_2$. Therefore, by induction, both factors on the right-hand side of (67) can be further decomposed into factors by rule (62). Thus, by the last statement about the components of $\mathfrak{D}_1$ and $\mathfrak{D}_2$, (67) becomes

$$\beta(\bar{\sigma} - \sigma_1 + \mathfrak{C}) = \prod_{\sigma \in \mathfrak{R}} (\bar{\sigma} - \sigma_1 + \mathfrak{C}_\sigma).$$

Now $\bar{\sigma}-\sigma_1$ can be cancelled on both sides, because, by ($27_2$), this signifies only the application of the same automorphism on both sides of this equation. Thus we have proved our decomposition formula (62).

Our next aim is the "coefficient-free" redrafting of (62) (where once again $\bar{f}(x)$ itself will occur, in place of its coefficients). We shall use the abbreviation

$$|\mathfrak{C}| = \sum_{\mu \in \mathfrak{C}} q_\mu \qquad (\mathfrak{C} \subseteqq \mathfrak{M}). \tag{75}$$

Clearly

$$\mathfrak{C} \to |\mathfrak{C}|$$

(cf. (39)) is a one-to-one mapping of the set of all complexes of $\mathfrak{M}$ onto the set $\mathscr{E}_m$. On the other hand, when we rewrite (40) using (75) we get

$$\beta(\mathfrak{C}) = \beta_{|\mathfrak{C}|} \qquad (\mathfrak{C} \subseteqq \mathfrak{M})$$

so that (25) becomes

$$\bar{f}(x) = \sum_{\mathfrak{C} \subseteqq \mathfrak{M}} \beta_{|\mathfrak{C}|} x^{|\mathfrak{C}|}. \tag{76}$$

We now introduce the partial sum of (76)

$$h(x) = \sum_{\mathfrak{C} \subseteqq \mathfrak{W}} \beta_{|\mathfrak{C}|} x^{|\mathfrak{C}|}. \tag{77}$$

It will be easy to express $\bar{f}(x)$ in terms of this "constituent" of it.

In order to do this, we remark that (75) implies the rule

$$|\mathfrak{A} \cup \mathfrak{B}| = |\mathfrak{A}| + |\mathfrak{B}| \qquad (\mathfrak{A}, \mathfrak{B} \subseteqq \mathfrak{M},\ \mathfrak{A} \cap \mathfrak{B} = \varnothing).$$

Repeated application of this to (63) gives

$$|\mathfrak{C}| = \sum_{\sigma \in \mathfrak{R}} |\mathfrak{C}_\sigma| \qquad (\mathfrak{C} \subseteqq \mathfrak{M}). \tag{78}$$

We see from the definition (61) (applied to all $\mathfrak{C} \subseteqq \mathfrak{M}$) that in (63) the components $\mathfrak{C}_\sigma$ run through all subsets of $\sigma + \mathfrak{W}$ independently of one another. Thus under application of formulae (62) and (78), formula (76) becomes

$$\bar{f}(x) = \prod_{\sigma \in \mathfrak{R}} \sum_{\mathfrak{K} \subseteqq \sigma + \mathfrak{W}} \beta(\mathfrak{K}) x^{|\mathfrak{K}|}. \tag{79}$$

For each $\sigma \in \mathfrak{R}$, however, the $\mathfrak{K} \subseteqq \sigma + \mathfrak{W}$ can be uniquely written in the form $\sigma + \mathfrak{C}$ ($\mathfrak{C} \subseteqq \mathfrak{W}$), showing that (79) can be transformed into

$$\bar{f}(x) = \prod_{\sigma \in \mathfrak{R}} \sum_{\mathfrak{C} \subseteqq \mathfrak{W}} \beta(\sigma + \mathfrak{C}) x^{|\sigma + \mathfrak{C}|}. \tag{80}$$

Since for elements $\sigma, \tau \in \mathfrak{M}$ with

$$\sigma \equiv \tau \pmod{\mathfrak{W}}$$

the $\sigma+\mathfrak{C}$ ($\mathfrak{C}\subseteqq\mathfrak{W}$) correspond, apart from the order in which they are taken, with the $\tau+\mathfrak{C}$ ($\mathfrak{C}\subseteqq\mathfrak{W}$), in (80) $\mathfrak{R}$ can be replaced by any other representative system $\mathfrak{R}'$ of $\mathfrak{M}$ mod $\mathfrak{W}$. We choose (cf. (55))

$$\mathfrak{R}' = \{0, -\varepsilon, -2\varepsilon, \ldots, -(e'-1)\varepsilon\}$$

for this purpose. Then (80) becomes

$$\bar{f}(x) = \prod_{k=0}^{e'-1} \sum_{\mathfrak{C}\subseteqq\mathfrak{W}} \beta(-k\varepsilon+\mathfrak{C})x^{|-k\varepsilon+\mathfrak{C}|},$$

whence after substitution from (41) we obtain

$$\bar{f}(x) = \prod_{k=0}^{e'-1} \sum_{\mathfrak{C}\subseteqq\mathfrak{W}} \beta(\mathfrak{C})^{P^k} x^{|-k\varepsilon+\mathfrak{C}|}. \tag{81}$$

In order to transform also the exponent of $x$, we note that by virtue of (17), (38) and (54), clearly

$$q_{-k\varepsilon+\mu} = P^k q_\mu \qquad (k=0, \ldots, e'-1;\ \mu\in\mathfrak{W})$$

holds. Thus by (75)

$$|-k\varepsilon+\mathfrak{C}| = P^k|\mathfrak{C}| \qquad (k=0, \ldots, e'-1;\ \mathfrak{C}\subseteqq\mathfrak{W}).$$

Substitution of this into (81) gives

$$\bar{f}(x) = \prod_{k=0}^{e'-1} \sum_{\mathfrak{C}\subseteqq\mathfrak{W}} \beta(\mathfrak{C})^{P^k} x^{P^k|\mathfrak{C}|} = \prod_{k=0}^{e'-1} \Big(\sum_{\mathfrak{C}\subseteqq\mathfrak{W}} \beta(\mathfrak{C})x^{|\mathfrak{C}|}\Big)^{P^k}.$$

Consequently, by (77), we obtain

$$\bar{f}(x)=h(x)^{\frac{P^{e'}-1}{P-1}}. \tag{82}$$

We now introduce the notation

$$Q=P^{e'} \tag{83}$$

and show that

$$h^{\circ Q}=1. \tag{84}$$

By (54) and (75) we have

$$|\mathfrak{W}| = \sum_{\mu\in\mathfrak{W}} q_\mu = \sum_{k=1}^{e} q_{ke'\varepsilon}.$$

Since, moreover, here (cf. (53)) we always have that $0<ke'\leqq m$, (38) allows us to write this last formula in the simpler form

$$|\mathfrak{W}| = \sum_{k=1}^{e} q_{ke'}.$$

According to (17), the terms $q_{ke'}$ are equal to

$$P^{m-ke'},$$

i.e. (cf. (53)) to

$$P^{e'(e-k)},$$

and thus, by (83), to

$$Q^{e-k}.$$

We therefore have that

$$|\mathfrak{W}| = \sum_{k-1}^{k} Q^{e-k} = Q^{e-1}+Q^{e-2}+\cdots+Q+1.$$

It follows from this that the partial sums on the right-hand side are just the $|\mathfrak{C}|$, formed with $\mathfrak{C} \subseteqq \mathfrak{W}$. In view of (77), this means that $h^{\circ}{}_{\varrho} \leqq 1$ holds. However, not only $\bar{f}(x)$ but also $h(x)$ has at least two terms, so that (84) must in fact be valid.

We extend this result to $f(x)$. By (23) we have

$$f(x) = x^{\frac{q-1}{P-1}} \bar{f}\left(\frac{1}{x}\right).$$

Thus, as a consequence of (82) and (83),

$$f(x) = x^{\frac{q-1}{P-1}} h\left(\frac{1}{x}\right)^{\frac{Q-1}{P-1}},$$

whence

$$f(x) = g(x)^{\frac{Q-1}{P-1}} \tag{85}$$

follows, with

$$g(x) = x^{\frac{q-1}{Q-1}} h\left(\frac{1}{x}\right). \tag{86}$$

Now, $(18_2)$ and (85) imply that $g(x) \in \mathsf{F}[x]$ and that

$$g^{\circ} = \frac{q-1}{Q-1}.$$

So, (84) and (86) clearly imply that

$$g^{\circ}{}_{\varrho} = 1. \tag{87}$$

Since by (53) $ee' = m$ and $e' > 1$, it follows from (16) and (83) that

$$P \underset{1}{|} Q \underset{1}{|} q, \qquad P < Q, \tag{88}$$

whereas (85), (87) and (88) show that $f(x)$ is non-primitive. Thus we have completed the proof of the "if" part of Theorem 12 for case (13).

In order to reduce the general case (11) to case (13), we put

$$f_\varrho(x) = f(x) - \varrho, \tag{89}$$

where $\varrho$ denotes the element appearing in (11). By Proposition 24, it follows from the assumption that $f_\varrho(x)$ is a solution of Problem III, belonging to the pair $P$, $\varphi(x+\varrho)$. The determinant of $f_\varrho(x)$ is evidently equal to

$$D(x) - \varrho f_{11}(x),$$

on account of (89), so that it vanishes because of (11). $f_\varrho(x)$ therefore belongs to the preceding case, so that it is a non-primitive solution of Problem III. Since, however,

$$f(x) = f_\varrho(x) + \varrho,$$

the same is valid also for $f(x)$. This proves Theorem 12 completely.

## § 28. On the primitive solutions of Problem III

We are now in a position to prove the following

THEOREM 13. *In order that at least one non-linear primitive solution of Problem III should belong to a given pair $P$, $\varphi(x)$, the marginal case $\varphi^\circ = \dfrac{P-1}{2}$ (so that $p \neq 2$) and $P^3 \leqq q$ must both arise.*

(The first of these necessary conditions asserts in other words that every non-linear primitive solution of Problem III is weak.)

We suppose that $f(x)$ is a non-linear primitive solution of Problem III, belonging to the given pair. As $f^\circ > 1$ and from Proposition 22, $P^2 \leqq q$ must hold. It also follows from Proposition 23 that the condition

$$f_1(x) \Big| \prod_{\bar\varphi(\varrho)=0} D_\varrho(x) \tag{1}$$

must be satisfied, where

$$\bar\varphi(x) = \frac{x^P - x}{\varphi(x)} \tag{2}$$

is the cofactor of $\varphi(x)$ and $D_\varrho(x)$ denotes the codeterminants of $f(x)$. According to the corollary of Theorem 12, these last are now not equal to 0:

$$D_\varrho(x) \neq 0 \qquad (\varrho \in \mathsf{F}_P,\ \bar\varphi(\varrho) = 0). \tag{3}$$

Apart from these we shall need for the proof nothing but the degree conditions

$$\varphi^\circ \geqq \frac{P-1}{2}, \quad f_1^\circ = \frac{q-1}{P-1}, \quad D_\varrho^\circ \leqq 2\,\frac{q_2-1}{P-1}, \tag{4}$$

which are likewise satisfied (partly by § 23).

By virtue of (3) the right-hand side of (1) is not equal to 0, so that its degree is at least as great as that of the left-hand side. By $(4_2)$ and $(4_3)$, this leads to the inequality

$$2\,\frac{q_2-1}{P-1}\,\bar{\varphi}^\circ \geqq \frac{q_1-1}{P-1}.$$

Hence, first of all, it follows that $q_2 > 1$, i.e. $P^3 \leqq q$, as stated above. It also follows that

$$\bar{\varphi}^\circ \geqq \frac{q-1}{2(q_2-1)} = \frac{P_{q_2}-1}{2(q_2-1)} = \frac{P}{2} + \frac{P-1}{2(q_2-1)}.$$

As the left-hand side is an integer, this gives

$$\bar{\varphi}^\circ \geqq \frac{P+1}{2}.$$

By (2), this can be replaced by

$$\varphi^\circ \leqq \frac{P-1}{2},$$

and therefore in $(4_1)$ the sign $=$ must hold. Thus Theorem 13 is proved.

## § 29. The solution of Problem III, apart from the marginal case

It is now easy to determine all the strong solutions of Problem III. By definition, the latter are those solutions, belonging to a pair $P$, $\varphi(x)$, for which

$$\frac{P-1}{2} < \varphi^\circ \leqq P-1 \tag{1}$$

holds. Therefore this will solve Problem III, except for the marginal case.

Let $f(x)$ be one of the solutions to be discussed. By Theorem 13, $f(x)$ is non-primitive, so that it possesses a successor $g(x)$, unless it is linear, which case, being trivial, will be disregarded for the time being. We infer from Theorem 9 that $g(x)$ is also a solution of Problem III and that it belongs to a pair $Q$, $\varphi(x)$ with

$$P \underset{1}{|} Q \underset{1}{|} q, \quad P < Q, \quad \psi(x) = \varphi\left(x^{\frac{Q-1}{P-1}} + \varrho\right), \quad \varrho \in \mathsf{F}_P.$$

As

$$\psi^\circ = \frac{Q-1}{P-1}\varphi^\circ > \frac{Q-1}{2}$$

follows from (1), $g(x)$ is again a strong solution. By repeated use of this result we establish the existence of a successor chain with initial term $f(x)$ and a linear last term; here not even the disregarded possibility of a linear $f(x)$ makes an exception. Thus, since $f(x)$ is therefore regular, we have

THEOREM 14. *The strong solutions of Problem III are regular.*

Since the regular solutions are known, by Theorem 10 (or 11), Problem III is actually solved by Theorem 14, apart from the marginal case. As it is, however, the latter case that primarily concerns us (cf. the Main Lemma), this will be the subject of the remaining two sections of this chapter.

## § 30. A part of the marginal case of Problem III without primitive solutions

In view of Theorem 14 we need investigate Problem III only in the marginal case. As already stated, it would be difficult to give a complete discussion of this marginal case, and we do not intend to do this. We shall, however, obtain results in this respect which will help us—as mentioned above—to solve our main problem, apart from a few exceptional cases (see Chapter V).

Of course, the most important question, as far as Problem III is concerned, is whether it possesses any primitive solutions that are not linear. It will be our aim in this and the following sections to prove that no (non-linear) primitive solutions exist belonging to certain pairs $P$, $\varphi(x)$. By Theorem 13, only the pairs $P$, $\varphi(y)$ with

$$P^3 \leqq q, \quad \varphi^\circ = \frac{P-1}{2} \qquad (p \neq 2) \tag{1}$$

are critical.

We shall prove in this section that Problem III has no primitive solutions belonging to the pair $P$, $\varphi(x)$, if apart from (1) we assume either

$$P \geqq 7, \quad P^3 = q, \quad \varphi^\dagger \geqq 3 \tag{2}$$

or

$$P \geqq 19, \quad P^4 \leqq q, \quad \varphi^\dagger \geqq 6. \tag{3}$$

The proof of this assertion will also be very difficult.

For greater clarity let us first assume only the validity of (1) (without (2) or (3)). We consider a primitive solution $f(x)$ of Problem III, belonging to the pair $P$, $\varphi(x)$. Our task is to deduce a contradiction in the cases (2), (3).

In connection with the polynomial $f(x)$ under consideration we shall take everything we need from § 23 without further reference. By our assumption the validity of (*inter alia*)

$$f(x) \in \mathsf{F}[x], \quad f^{\circ} = \frac{q-1}{P-1}, \quad f^{\circ_P} = 1 \tag{4}$$

follows as well as that of the necessary condition

$$f_1(x) \,\Big|\, \prod_{\bar{\varphi}(\varrho)=0} D_\varrho(x), \tag{5}$$

obtained in Proposition 23, where $\varrho$ runs through the zeros lying in $\mathsf{F}_P$ of the cofactor

$$\bar{\varphi}(x) = \frac{x^P - x}{\varphi(x)} \tag{6}$$

of $\varphi(x)$, the corresponding

$$D_\varrho(x) = D(x) - \varrho f_{11}(x)$$

being all the codeterminants of $f(x)$. From the corollary of Theorem 12, the latter are not equal to 0, and so the same holds for the right-hand side of (5) as well.

It is best to write $\bar{\varphi}(x)$ straight away in its coefficient representation

$$\bar{\varphi}(x) = \sum_{i=0}^{\frac{P+1}{2}} \varkappa_i x^{\frac{P+1}{2}-i} \qquad (\varkappa_i \in \mathsf{F}_P;\ \varkappa_0 = 1) \tag{7}$$

(already introduced above but not yet used). Thus after multiplying out as indicated (5) assumes the form

$$f_1(x) \,\Big|\, \sum_{i=0}^{\frac{P+1}{2}} \varkappa_i D(x)^{\frac{P+1}{2}-i} f_{11}(x)^i \tag{8}$$

which we are familiar with from Proposition 23, in which—as we must repeatedly point out—the right-hand side does not vanish.

NOTE 1. Although the divisibilities

$$x^q - x \,|\, f(x)^P - f(x), \quad \varphi\big(f(x)\big) \,\big|\, x^q - x,$$

also follow from the assumption, we shall disregard these in what follows, basing our procedure on (4) and (8). (Or, briefly: only the consequence (8) of these divisibilities will be used.)

Of course condition (8) is too complicated to be treated "directly", but with the aid of a simple device we shall infer the lacunarity of the right-hand side from that of the left-hand side, and this will enable us to proceed. This device consists simply in changing (8) with the aid of a (uniquely determined) polynomial $g(x) \in \mathsf{F}[x]$ into the equation

$$f_1(x)g(x) = \sum_{i=0}^{\frac{P+1}{2}} \varkappa_i D(x)^{\frac{P+1}{2}-i} f_{11}(x)^i. \tag{9}$$

We shall call $g(x)$ the *catalyst* of $f(x)$ (with respect to $\varphi(x)$). This nomenclature, borrowed from chemistry, was chosen because $g(x)$ will play a brief but vital role in the simplification of our problem, although $g(x)$ will remain almost completely unexplored, except for an upper estimation of its degree in (9).

Since

$$f_1^\circ = \frac{q_1-1}{P-1}, \quad f_{11}^\circ = \frac{q_2-1}{P-1}, \quad D^\circ \leqq 2\frac{q_2-1}{P-1} (= 2f_{11}^\circ), \tag{10}$$

it follows from (9) that

$$g^\circ \leqq (P+1)\frac{q_2-1}{P-1} - \frac{q_1-1}{P-1},$$

i.e. (because $q_1 = P_{q_2}$)

$$g^\circ \leqq \frac{q_2-1}{P-1} - 1 (= f_{11}^\circ - 1). \tag{11}$$

The use of (11) to prove lacunarity of the right-hand side of (8) (or of (9)) will be made possible by exploiting the lacunarity of $f_1(x)$, which follows from $(4_2)$ and $(4_3)$. This lacunarity is reflected in the decomposition

$$f_1(x) = f_{10}(x) + x^{q_2} f_{11}(x) \tag{12}$$

and in the degree estimations

$$f_{10}^\circ \leqq f_{11}^\circ \left(= \frac{q_2-1}{P-1}\right), \tag{13}$$

according to which $f_1(x)$ is lacunary at least from $\frac{q_2-1}{P-1}$ up to $q_2$. (Of course, if $P^4 \geqq q$ $f_1(x)$ has further gaps, but these will not be considered here.)

Using the coefficient operator $\langle\ \rangle_k$, this lacunarity can be expressed in the form

$$\langle f_1(x)\rangle_k = 0 \quad \left(\frac{q_2-1}{P-1} + 1 \leqq k \leqq q_2 - 1\right),$$

whence, by (11), it follows that

$$\langle f_1(x)g(x)\rangle_k = 0 \quad \left(2\frac{q_2-1}{P-1} \leqq k \leqq q_2-1\right).$$

Consequently, by (9) we have

$$\left\langle \sum_{i=0}^{\frac{P+1}{2}} \varkappa_i D(x)^{\frac{P+1}{2}-i} f_{11}(x)^i \right\rangle_k = 0 \quad \left(2\frac{q_2-1}{P-1} \leqq k \leqq q_2-1\right). \tag{14}$$

(The catalyst $g(x)$ will not be used any more.)

We agree to write as $R(x)$, $R_1(x)$, ... polynomials from $\mathsf{F}[x]$ the degrees of which we shall estimate from above, all their other features being completely disregarded. Then (14) may be written in the form

$$\sum_{i=0}^{\frac{P+1}{2}} \varkappa_i D(x)^{\frac{P+1}{2}-i} f_{11}(x)^i \equiv R(x) \pmod{x^{q_2}} \quad \left(R^\circ \leqq 2\frac{q_2-1}{P-1}-1\right), \tag{15}$$

which conforms better to our purpose.

Before applying (15) we must obtain some preliminary results. We first endeavour to estimate the degree $D^\circ$ from below. For this purpose we go back to divisibility (8). As its right-hand side is not 0, its degree is at least as great as that of the left-hand side. It follows from this that at least one of the inequalities

$$\frac{P+1}{2} D^\circ \geqq f_1^\circ, \quad \frac{P+1}{2} f_{11}^\circ \geqq f_1^\circ$$

must hold. The second is clearly false, by (10), $(10_2)$ and $q_1 = Pq_2$. Therefore the first must be valid. All this, in view of (10), proves the validity of

$$D^\circ \geqq 2\frac{q_1-1}{P^2-1} \, (> f_{11}^\circ). \tag{16}$$

We must also decompose $D(x)$ into suitable additive "constituents". By the definition

$$D(x) = \begin{vmatrix} f_{00}(x) & f_{01}(x) \\ f_{10}(x) & f_{11}(x) \end{vmatrix}, \tag{17}$$

whence, as $f_{ij}^{\circ_P} \leqq 1$ $(i, j = 0, 1)$ we obtain

$$D^{\circ_P} \leqq 2. \tag{18}$$

Thus from $(10_3)$ the existence of the decomposition

$$D(x) = D_0(x) + x^{q_3} D_1(x) + x^{2q_3} D_2(x) \tag{19}$$

follows with

$$D_i(x) \in \mathsf{F}[x], \quad D_i^\circ \leqq 2\frac{q_3-1}{P-1}, \quad D_i^{\circ_P} \leqq 2 \qquad (i = 0, 1, 2). \tag{20}$$

We at once remark that if $P^4 \leqq q$ the $D_i(x)$ can be further decomposed on the strength of (20) in a similar way; we shall, however, be interested in this decomposition only for $i=2$. Accordingly we obtain

$$D_2(x) = E_0(x) + x^{q_4} E_1(x) + x^{2q_4} E_2(x) \qquad (\text{if } P^4 \leqq q) \tag{21}$$

with

$$E_i(x) \in \mathsf{F}[x], \quad E_i^\circ \leqq 2\frac{q_4-1}{P-1} \qquad (i = 0, 1, 2). \tag{22}$$

We note that from (16), (19) and from the case $i=2$ of $(20_2)$ (since $q_1 = P^2 q_3$) the estimation

$$D_2^\circ \geqq 2\frac{q_3-1}{P-1} \tag{23}$$

follows, whence

$$D_2(x) \neq 0. \tag{24}$$

(As a consequence of $(20_2)$ the inequality (23) is also true with $=$, a point that will not, however, concern us.)

The last and particularly important preliminary result is to prove that

$$D_1(x)^2 - 4D_0(x)D_2(x) \neq 0 \qquad (\text{if } \varphi^\downarrow \geqq 2). \tag{25}$$

(It is easy to see that (25) remains valid without the restriction $\varphi^\downarrow \geqq 2$, but we shall not waste our time on this, as we shall need (25) only in case (3).)

To prove (25) we shall assume that the left-hand side of (25) vanishes. Then by (19)

$$D_0(x)D(x) = \left(D_0(x) + \frac{1}{2}x^{q_3}D_1(x)\right)^2,$$

so that (as $Pq_3 = q_2$)

$$(D_0(x)D(x))^{\frac{P+1}{2}} = \left(D_0(x) + \frac{1}{2}x^{q_3}D_1(x)\right)^{P+1} =$$

$$= D_0(x)^{P+1} + \frac{1}{2}x^{q_3}D_0(x)^P D_1(x) + \frac{1}{2}x^{q_2}D_0(x)^P D_1(x)^P + \frac{1}{4}x^{q_2+q_3}D_1(x)^{P+1}.$$

In the last two terms we must consider the equations

$$D_1(x)^{P-1} = D_0(x)^{\frac{P-1}{2}} D_2(x)^{\frac{P-1}{2}}$$

and

$$D_1(x)^{P+1} = 4D_0(x)^{\frac{P+1}{2}} D_2(x)^{\frac{P+1}{2}}$$

which follow from our assumption [as $2^{P-1} \equiv 1 \pmod{p}$]. Then if $D_0(x) \neq 0$ after a reduction we obtain the equation

$$D(x)^{\frac{P+1}{2}} = D_0(x)^{\frac{P+1}{2}} + \frac{1}{2} x^{q_3} D_0(x)^{\frac{P-1}{2}} D_1(x) + $$
$$+ \frac{1}{2} x^{q_2} D_1(x) D_2(x)^{\frac{P-1}{2}} + x^{q_2+q_3} D_2(x)^{\frac{P+1}{2}}.$$

The disregarded case $D_0(x)=0$ is not an exception, since the assumption implies that then $D_1(x)=0$ holds, whence

$$D(x) = x^{2q_3} D_2(x)$$

and thus (because of $Pq_3 = q_2$) it follows that the preceding equation is trivially satisfied.

The congruence

$$D(x)^{\frac{P+1}{2}} \equiv D_0(x)^{\frac{P+1}{2}} + \frac{1}{2} x^{q_3} D_0(x)^{\frac{P-1}{2}} D_1(x) \qquad (\operatorname{mod} x^{q_2})$$

holds *a fortiori*. The degree of the right-hand side is by $(20_2)$ at most equal to

$$q_3 + (P+1)\frac{q_3-1}{P-1} \left( = 2q_3 + 2\frac{q_3-1}{P-1} - 1 = 2\frac{q_2-1}{P-1} - 1 \right).$$

It follows from this that in (15) the term belonging to $i=0$ can be cancelled (by passing over to a new $R(x)$ with an unchanged degree estimation).

As $\bar{\varphi}^{\downarrow} \geqq 2$ clearly follows from $\varphi^{\downarrow} \geqq 2$, we now have that $\varkappa_1 = 0$. Since, however, $\bar{\varphi}(x)$ is not a monomial, it follows that a $k$ exists, such that

$$\varkappa_1 = \cdots = \varkappa_{k-1} = 0, \quad \varkappa_k \neq 0 \quad \left( 2 \leqq k \leqq \frac{P+1}{2} \right);$$

and therefore (15) turns into

$$\sum_{i=k}^{\frac{P+1}{2}} \varkappa_i D(x)^{\frac{P+1}{2}-i} f_{11}(x)^i \equiv R(x) \quad (\operatorname{mod} x^{q_2}) \quad \left( R^{\circ} \leqq 2\frac{q_3-1}{P-1} - 1 \right).$$

Since by (16) we have $D^{\circ} > f_{11}^{\circ}$, the degree of the left-hand side of this congruence is equal to

$$\left( \frac{P+1}{2} - k \right) D^{\circ} + k f_{11}^{\circ}.$$

This number is, by $(10_2)$ and $(10_3)$, at most equal to

$$(P+1-k) f_{11}^{\circ} \left( \leqq (P-1) f_{11}^{\circ} = q_2 - 1 \right).$$

Therefore

$$\left(\frac{P+1}{2}-k\right) D^{\circ}+kf_{11}^{\varphi} \leqq 2\frac{q_2-1}{P-1}-1$$

must hold, whence, as $D^{\circ}>f_{11}^{\varphi}$ and $k \leqq \dfrac{P+1}{2}$,

$$kf_{11}^{\varphi} < 2\frac{q_2-1}{P-1},$$

i.e. (by $(10_2)$) $k<2$ follovs. This contradiction proves (25).

We shall now solve the simple case (2) separately, as it would otherwise be a nuisance later on. From $(10_3)$ and (16), we now have $D^{\circ}=2$. By Proposition 29, $f(x)$ may be replaced by any $f(x+\xi)$ $(\xi \in \mathsf{F})$, and thus, by virtue of Propositions 2 and 26 we are allowed to assume that

$$D^{\downarrow} \geqq 2.$$

Thus

$$D_2(x) = \alpha x^2+\beta \qquad (\alpha \in \mathsf{F}\setminus 0,\ \beta \in \mathsf{F})$$

holds for suitable $\alpha$, $\beta$. Moreover,

$$f_{11}(x) = x+\gamma \qquad (\gamma \in \mathsf{F})$$

is now valid for suitable $\gamma$. Now from (2), $(2_3)$ and (6) we can infer that $\bar{\varphi}^{\downarrow} \geqq 3$. Since $\bar{\varphi}(x)$ is not a monomial, the existence of a $k$ such that

$$\varkappa_0 = 1, \quad \varkappa_1 = \cdots = \varkappa_{k-1} = 0, \quad \varkappa_k \neq 0 \quad \left(3 \leqq k \leqq \frac{P+1}{2}\right)$$

follows (cf. (7)).

The degree of the left-hand side of (15) is now $P+1$ $(>P=q_2)$. If $\beta \neq 0$ its second degree is equal to $P-1$ and if $\beta=0$ to

$$2\left(\frac{P+1}{2}-k\right)+k(=P+1-k);$$

in both cases it lies between $\dfrac{P+1}{2}$ and $P-1$. On the other hand, we now have $R^{\circ} \leqq 1$. This obvious contradiction proves that no $f(x)$ exists for case (2).

We now assume the validity of (3). In view of $(10_2)$ and $(10_3)$, the degree of the summand in (15) is at most equal to

$$\left(\frac{P+1}{2}-i\right) D^{\circ}+if_{11}^{\varphi} \leqq (P+1-i)f_{11} = (P+1-i)\frac{q_2-1}{P-1}.$$

We must, however, bear in mind that on account of ($3_3$) and (6), $\varphi^{\dagger} \geqq 6$, i.e. (cf. (7))

$$x_1 = \cdots = x_5 = 0.$$

Hence it follows that the degree of the part of the sum on the left-hand side of (15) that belongs to $i \neq 0$ is at most equal to

$$(P-5)\frac{q_2-1}{P+1}\left(= q_2 - 4\frac{q_2-1}{P-1} - 1\right).$$

This number is (because of ($3_1$)) greater than the degree of the right-hand side of (15); so that

$$D(x)^{\frac{P+1}{2}} \equiv R(x) \pmod{x^{q_2}} \quad \left(R^{\circ} \leqq q_2 - 4\frac{q_2-1}{P-1} - 1\right) \tag{26}$$

follows from (15) with a new $R(x)$. (We shall not use (15) any more since its weaker form (26) suffices for our purpose.)

By (19), equation (26) becomes

$$\sum_{i+j+k=\frac{P+1}{2}} \begin{pmatrix} \frac{P+1}{2} \\ i, j, k \end{pmatrix} x^{(j+2k)q_3} D_0(x)^i D_1(x)^j D_2(x)^k \equiv R(x) \pmod{x^{q_2}} \tag{27}$$

$$\left(R^{\circ} \leqq q_2 - 4\frac{q_2-1}{P-1} - 1\right),$$

where the polynomial coefficient

$$\begin{pmatrix} \frac{P+1}{2} \\ i, j, k \end{pmatrix} = \frac{\frac{P+1}{2}!}{i!\,j!\,k!} \tag{28}$$

has been applied. (We treat the latter as an element of $\mathsf{F}_p$.)

We take an

$$l = 1, 2, 3, 4. \tag{29}$$

We split the left-hand side of (27) into two partial sums I and II which come about by retaining only the terms with

$$j + 2k \geqq P - l \quad \text{or} \quad j + 2k \leqq P - l - 1,$$

respectively. Then (27) can be written in the form

$$I = -II + R(x) \pmod{x^{q_2}} \qquad \left(R^{\circ} \leqq q_2 - 4\frac{q_2-1}{P-1} - 1\right).$$

The degree of II is at most equal to

$$(P-l-1)q_3+(P+1)\frac{q_3-1}{P-1}\left(=(P-l)q_3+2\frac{q_3-1}{P-1}-1\right),$$

because of $(20_2)$. On the other hand, as $q_2=Pq_3$, we have

$$R^{\circ} \leqq Pq_3-4q_3-4\frac{q_3-1}{P-1}-1.$$

Since $l\leqq 4$, the right-hand side is here less than the previously calculated upper bound; therefore

$$I \equiv R_l(x) \pmod{x^{q_2}} \qquad \left(R_l^{\circ} \leqq (P-1)q_3+2\frac{q_3-1}{P-1}-1\right)$$

with suitable $R_1(l=1, \ldots, 4)$. In detail

$$\sum_{\substack{i+j+k=\frac{P+1}{2}\\ j+2k\geqq P-l}} \left(\begin{matrix}\frac{P+1}{2}\\ i, j, k\end{matrix}\right) x^{(j+2k)q_3} D_0(x)^i D_1(x)^j D_2(x)^k \equiv R_l(x) \pmod{x^{q_2}}.$$

The left-hand side is divisible by

$$x^{(P-l)q_3}(=x^{q_2-lq_3}),$$

and thus the same also holds for $R_l(x)$. After division by this power of $\varkappa$ and using $j+2k = P+1-2i-j$, we get

$$\sum_{\substack{i+j+k=\frac{P+1}{2}\\ 2i+j\leqq l+1}} \left(\begin{matrix}\frac{P+1}{2}\\ i, j, k\end{matrix}\right) x^{(l+1-2i-j)q_3} D_0(x)^i D_1(x)^j D_2(x)^k \equiv R_l(x) \pmod{x^{lq_3}}$$

$$\left(R_l^{\circ} \leqq 2\frac{q_3-1}{P-1}-1\right),$$

with a new $R_l(x)$. The terms with $2i+j \leqq 1$ can be cancelled, as they are divisible by $x^{lq_3}$, so that only those with $2i+j \geqq 2$ are retained. Applying (28), we thus have

$$\sum_{2\leqq 2i+j\leqq l+1} \binom{\frac{1}{2}}{i+j}\binom{i+j}{i} x^{(l+1-2i-j)q_3} D_0(x)^i D_1(x)^j D_2(x)^{\frac{P+1}{2}-i-j} \equiv R_l(x) \pmod{x^{lq_3}}.$$

For convenience we now introduce the notation

$$F_k(x) = \sum_{2i+j=k} \binom{\frac{1}{2}}{i+j}\binom{i+j}{i} D_0(x)^i D_1(x)^j D_2(x)^i \qquad (k = 2, 3, 4, 5). \quad (30)$$

Our result then takes the form

$$\sum_{k=2}^{l+1} x^{(l+1-k)q_3} D_2(x)^{\frac{P+1}{2}-k} F_k(x) \equiv R_P(x) \qquad (\mathrm{mod}\ x^{lq_3}) \tag{31}$$

$$\left(l = 1, 2, 3, 4;\ R_l^{\circ} \leqq 2\frac{q_3-1}{P-1}-1\right).$$

In preparation for the application of this formula we calculate from (30) that:

$$F_2(x) = -\frac{1}{8} D_1(x)^2 + \frac{1}{2} D_0(x) D_2(x),$$

$$F_3(x) = \frac{1}{16} D_1(x)^3 - \frac{1}{4} D_0(x) D_1(x) D_2(x),$$

$$F_4(x) = -\frac{5}{128} D_1(x)^4 + \frac{3}{16} D_0(x) D_1(x)^2 D_2(x) - \frac{1}{8} D_0(x)^2 D_2(x)^2,$$

$$F_5(x) = \frac{7}{256} D_1(x)^5 - \frac{5}{32} D_0(x) D_1(x)^3 D_2(x) + \frac{3}{16} D_0(x)^2 D_1(x) D_2(x)^2.$$

For brevity we write (cf. (25))

$$\Omega(x) = \Omega_1(x) = D_1(x)^2 - 4D_0(x) D_2(x) \quad (\neq 0), \tag{32}$$

$$\Omega_2(x) = 5D_1(x)^2 - 4D_0(x) D_2(x), \tag{33}$$

$$\Omega_3(x) = 7D_1(x)^2 - 12D_0(x) D_2(x). \tag{34}$$

Then we have

$$F_2(x) = -\frac{1}{8}\Omega(x) \quad (\neq 0), \tag{35}$$

$$F_3(x) = \frac{1}{16} D_1(x)\Omega(x), \tag{36}$$

$$F_4(x) = -\frac{1}{128}\Omega(x)\Omega_2(x), \tag{37}$$

$$F_5(x) = \frac{1}{256} D_1(x)\Omega(x)\Omega_3(x). \tag{38}$$

Note also that

$$\Omega_i^{\circ} \leqq 4\frac{q_3-1}{P-1} \qquad (i = 1, 2, 3) \tag{39}$$

holds by $(20_2)$ and (32)—(34). Moreover,

$$F_k^{\circ} \leqq 2k\frac{q_3-1}{P-1} \qquad (k = 2, 3, 4, 5) \tag{40}$$

follows from $(20_2)$ and (30). The application of the congruences (31) will be possible partly as a result of the happy chance that we are able to deduce certain equations from them (admittedly with the aid of certain simple devices). Let us write the left-hand side of (31) in the form

$$D_2(x)^{\frac{P-1}{2}-l} \sum_{k=2}^{l+1} x^{(l+1-k)q_3} D_2(x)^{l+1-k} F_k(x).$$

Then, by raising (31) to the second power and then multiplying it by

$$D_2(x)^{2l+1}$$

we obtain the congruence

$$D_2(x)^P \left( \sum_{k=2}^{l+1} x^{(l+1-k)q_3} D_2(x)^{l+1-k} F_k(x) \right)^2 \equiv D_2(x)^{2l+1} R_l(x)^2 \pmod{x^{lq_3}}.$$

On the other hand, by (21)

$$D_2(x)^P = E_0(x)^P + x^{q_3} E_1(x)^P + x^{2q_3} E_2(x)^P$$

(because $Pq_4 = q_3$).

Keeping both these results in view, let us form the polynomials

$$\Phi_l(x, y) = \left(E_0(x)^P + yE_1(x)^P + y^2 E_2(x)^P\right) \left( \sum_{k=2}^{l+1} y^{l+1-k} D_2(x)^{l+1-k} F_k(x) \right)^2 \tag{41}$$

$$(l = 1, 2, 3, 4)$$

using a second indeterminate $y$; then we get

$$\Phi_l(x, x^{q_3}) \equiv D_2(x)^{2l+1} R_l(x)^2 \pmod{x^{lq_3}} \tag{42}$$

$$\left( l = 1, 2, 3, 4;\ R_l^{\circ} \leqq 2\frac{q_3 - 1}{P-1} - 1 \right).$$

So we arrive at the desired equations as follows. We see from (41) that $\Phi_l(x, y)$ is of degree $\leqq 2l$ in $y$. We therefore put

$$\Phi_l(x, y) = \sum_{j=0}^{2l} y^j \Phi_{lj}(x) \qquad (l = 1, 2, 3, 4;\ \Phi_{lj}(x) \in \mathsf{F}[x]). \tag{43}$$

Now by $(22_2)$

$$(E_i(x)^P)^{\circ} \leqq 2P\frac{q_4 - 1}{P-1} = 2\frac{q_3 - 1}{P-1} - 2 \qquad (i = 0, 1, 2).$$

Furthermore, $(20_2)$ and (40) imply that

$$(D_2(x)^{l+1-k} F_k(x))^{\circ} \leqq 2(l+1)\frac{q_3 - 1}{P-1} - 2 \qquad (l = 1, 2, 3, 4;\ 2 \leqq k \leqq l+1).$$

From these two inequalities the degree estimation

$$\Phi_{lj}^{\circ} \leqq 2\frac{q_3-1}{P-1} - 2 + 4(l+1)\frac{q_3-1}{P-1} \left( = (4l+6)\frac{q_3-1}{P-1} - 2 \right)$$

$$(l = 1, 2, 3, 4;\ j = 0, \ldots, 2l)$$

follows by (41) and (43). It may be easily seen that the upper bound thus obtained is less than $q_3$ if $l=1, 2, 3$, by (3). On the other hand, the degree of the right-hand side of (42) is, from ($20_2$), at most equal to

$$2(2l+1)\frac{q_3-1}{P-1} + 4\frac{q_3-1}{P-1} - 2 \left( = (4l+6)\frac{q_3-1}{P-1} - 2 \right).$$

This upper bound agrees with the last one (and thus for $l=1, 2, 3$ it is likewise less than $q_3$). Therefore (42) evidently implies the equations

$$\Phi_{l0}(x) = D_2(x)^{2l+1} R_l(x)^2, \quad \Phi_{lj}(x) = 0 \qquad (l = 1, 2, 3;\ j = 1, \ldots, l-1), \tag{44}$$

as is seen by expressing the left-hand side of the congruence (42) with the help of (43) in the form

$$\sum_{j=0}^{2l} x^{jq_3} \Phi_{lj}(x)$$

(where, of course, only the terms with $j=0, \ldots, l-1$ come into consideration, because all the others are divisible by $x^{lq_3}$) and comparing the coefficients on both sides with one another. These are the equations that were promised above.

[We see from the congruences (31), that apart from (44) no further equations can be obtained in this way. The case $l=4$ of (31) has not been used at all for the deduction of equations (44). Although the validity of (44) for $P \geqq 23$ can be derived from the case $l=4$ of (31) with an identical proof, we shall ignore this, as we shall later draw certain conclusions from the case $l=4$ of (31) in a different way, these conclusions being valid for all $P \geqq 19$.]

If $l=1$, (44) consists of the single equation

$$\Phi_{10}(x) = D_2(x)^3 R_1(x)^2.$$

By (41) and (43) the left-hand side is equal to

$$E_0(x)^P F_2(x)^2,$$

whence by (35) we obtain the equation

$$E_0(x)^P \Omega(x)^2 = 64 D_2(x)^3 R_1(x)^2. \tag{45}$$

An equation of the form

$$\cdots = \square$$

denotes the fact that the left-hand side is a square in $\mathsf{F}[x]$.

In view of $\Omega(x) \neq 0$ (see (32)) it follows from (45) that

$$E_0(x) D_2(x) = \square .$$

After substituting from (21) we have

$$E_0(x)\big(E_0(x) + x^{q_4} E_1(x) + (x)^{2q_4} E_2(x)\big) = \square .$$

In order to be able to use this relation we shall need

PROPOSITION 35. *If*

$$a(x)^2 + x^k b(x) + x^{2k} c(x) = \square \tag{47}$$

*is valid for a natural number k and for*

$$a(x), b(x), c(x) \in \mathsf{F}[x]; \quad x \nmid a(x); \quad b^\circ < k; \quad 2a^\circ + c^\circ < 2k, \tag{48}$$

*then*

$$b(x)^2 - 4a(x)^2 c(x) = 0 \tag{49}$$

*holds.*

For, according to our assumption there exists an $u(x) \in \mathsf{F}[x]$ such that

$$a(x)^2 + x^k b(x) + x^{2k} c(x) = \big(a(x) + u(x)\big)^2. \tag{50}$$

We now show that the divisibility

$$x^k | u(x) \tag{51}$$

holds for suitable $u(x)$.

(50) implies that

$$x^k | 2a(x) + u(x).$$

By $(48_2)$ and $(48_3)$ one of the factors is prime to $x$, so that the other must be divisible by $x^k$. If, therefore, (51) is false, then

$$x^k | u(x)\big(2a(x) + u(x)\big) \tag{52}$$

is true. Now in (50) $u(x)$ can be replaced by

$$v(x) = -2a(x) - u(x).$$

Since, moreover, $x^k | v(x)$ is satisfied by (52), we have established that (51) can be satisfied.

We write the $u(x)$ thus found in the form

$$u(x) = x^k w(x) \qquad \big(w(x) \in \mathsf{F}[x]\big). \tag{53}$$

Substituting into (50) and dividing by $x^k$ we get

$$b(x)+x^k c(x) = 2a(x)w(x)+x^k w(x)^2 (=w(x)(2a(x)+x^k w(x))). \tag{54}$$

If $c(x)=0$ holds, then the left-hand side of (54) is of degree $<k$, as a consequence of $(48_3)$. On the other hand, by virtue of $(48_4)$ we have $a^\circ<k$, and we therefore obtain $w(x)=0$. As, however, in this case (54) also implies that $b(x)=0$, (49) is satisfied in this case.

If $c(x)\neq 0$, then, again by virtue of $(48_3)$, the left-hand side of (54) is of degree $k+c^\circ$, while the right-hand side is again of degree $k+2w^\circ$ because $a^\circ<k$, whence it follows that

$$c^\circ=2w^\circ.$$

Therefore, using $(48_4)$, we have

$$a^\circ+w^\circ < k.$$

Because of this inequality and of $(48_3)$, (54) splits into the system of equations

$$b(x)=2a(x)w(x), \quad c(x)=w(x)^2.$$

The elimination of $w(x)$ leads to (49). Thus we have proved Proposition 35.

We return to (46). When $E_0(x)\neq 0$ we define $e$ by

$$x^e \| E_0(x)$$

and note that, by $(22_2)$,

$$e<q_4.$$

After dividing (46) by $x^{2e}$ we get

$$\left(\frac{E_0(x)}{x^e}\right)^2 + x^{q_4-e}\frac{E_0(x)}{x^e}E_1(x)+x^{2(q_4-e)}E_0(x)E_2(x)=\square.$$

It is therefore clear that the premises of Proposition 35 are now satisfied for $k=q_4-e$ and

$$a(x)=\frac{E_0(x)}{x^e}, \quad b(x)=\frac{E_0(x)}{x^e}E_1(x), \quad c(x)=E_0(x)E_2(x).$$

It follows that for these polynomials equation (49) holds. After multiplying by $x^{2e}$, this gives the equation

$$E_0(x)^2E_1(x)^2-4E_0(x)^3E_2(x)=0,$$

which is clearly equivalent to

$$E_0(x)(E_1(x)^2-4E_0(x)E_2(x))=0. \tag{55}$$

(It will be seen below that the second factor vanishes.)

If $l=2$, (44) consists of the two equations

$$\Phi_{20}(x)=D_2(x)^5 R_2(x); \quad \Phi_{21}(x)=0.$$

From (41) and (43), the left-hand sides are found to be

$$E_0(x)^P F_3(x)^2 \quad \text{and} \quad 2E_0(x)^P D_2(x) F_2(x) F_3(x)+E_1(x)^P F_3(x)^2,$$

respectively, so that substituting (35) and (36) (and multiplying by 256 and $-256$, respectively) we get

$$E_0(x)^P D_1(x)^2 \Omega(x)^2=256 D_2(x)^5 R_2(x)^2$$

and

$$(4E_0(x)^P D_2(x)-E_1(x)^P D_1(x)) D_1(x) \Omega(x)^2 = 0.$$

(It was only for completeness that we wrote down the first of these equations, as, compared with (45), it affords us no new information that could be used in what follows.) As $\Omega(x)\neq 0$ (cf. (32)) the last factor in the second equation can be cancelled, giving

$$(4E_0(x)^P D_2(x)-E_1(x)^P D_1(x)) D_1(x) = 0. \tag{56}$$

We shall prove that

$$D_1(x)\neq 0, \tag{57}$$

whence it follows that the first factor in (56) vanishes, i.e. that

$$4E_0(x)^P D_2(x)=E_1(x)^P D_1(x) \tag{58}$$

holds.

We assume $D_1(x)=0$, in order to produce a contradiction. To do this, we must go back to congruence (26). By (19), the latter now runs as follows:

$$(D_0(x)+x^{2q_3} D_2(x))^{\frac{P+1}{2}} \equiv R(x) \pmod{x^{q_2}} \quad \left(R^\circ \leqq q_2-4\frac{q_2-1}{P-1}-1\right).$$

Hence

$$\sum_{i=0}^{\frac{P+1}{2}} \binom{\frac{1}{2}}{i} x^{2iq_3} D_2(x)^i D_0(x)^{\frac{P+1}{2}-i} \equiv R(x) \pmod{x^{q_3}}.$$

From ($20_2$), the summand contains only terms whose degree lies between $2iq_3$ and

$$2iq_3+(P+1)\frac{q_3-1}{P-1}\,(< 2(i+1)q_3),$$

consequently the exponent sets of the summands are pairwise disjoint (i.e.

the sum itself is direct). In particular, the summand belonging to $i = \frac{P-3}{2}$ is equal to

$$-\frac{1}{8}x^{q_2-3q_3}D_2(x)^{\frac{P-3}{2}}D_0(x)^2$$

which does not vanish, since by (25) $\big($and $D_1(x)=0\big)$ $D_0(x)\neq 0$ and $D_2(x)\neq 0$ and therefore its degree is at least equal to $q_2-3q_3$, which is impossible, however, as this number is obviously greater than $R^{\circ}$. The contradiction thus obtained proves (57).

We are now in position to replace (55) as promised by the stronger equation

$$E_1(x)^2-4E_0(x)E_2(x) = 0. \tag{59}$$

If $E_0(x)\neq 0$ (59) follows from (55). If $E_0(x)=0$ the validity of $E_1(x)=0$ follows from (57) and (58). Thus (59) is proved.

If $l=3$, (44) consists of the three equations

$$\Phi_{30}(x)=D_2(x)^7R_3(x)^2,\quad \Phi_{31}(x)=0,\quad \Phi_{32}(x)=0.$$

After calculating the left-hand sides from (41) and (43) we have the three equations

$$E_0(x)^PF_4(x)^2=D_2(x)^7R_3(x)^2,$$

$$2E_0(x)^PD_2(x)F_3(x)F_4(x)+E_1(x)^PF_4(x)^2 = 0, \tag{60}$$

and

$$E_0(x)^P\big(2D_2(x)^2F_2(x)F_4(x)+D_2(x)^2F_3(x)^2\big)+ \tag{61}$$
$$+2E_1(x)^PD_2(x)F_3(x)F_4(x)+E_2(x)^PF_4(x)^2 = 0.$$

(In what follows we shall use only the last two equations, as we can draw no further inferences from the first.)

From (60) we obtain after multiplying by 2 and substituting from (58)

$$E_1(x)^P\big(D_1(x)F_3(x)F_4(x)+2F_4(x)^2\big) = 0.$$

This equation is equivalent to

$$E_1(x)F_4(x)\big(D_1(x)F_3(x)+2F_4(x)\big) = 0.$$

We multiply by $-128\cdot 64$ and substitute (36), (37):

$$E_1(x)\Omega(x)\Omega_2(x)\big(4D_1(x)^2\Omega(x)-\Omega_2(x)\Omega(x)\big) = 0.$$

As, by (32), (33), we have that

$$4D_1(x)^2-\Omega_2(x) = -\Omega(x),$$

then after substitution we obtain

$$E_1(x)\,\Omega(x)^3\,\Omega_2(x)=0,$$

so that (by $\Omega(x)\neq 0$) simply

$$E_1(x)\,\Omega_2(x)=0. \tag{62}$$

We now prove that this can be replaced by the stronger assertion

$$\Omega_2(x)=0. \tag{63}$$

For this purpose we assume that

$$\Omega_2(x)\neq 0.$$

Then, from (62),

$$E_1(x)=0.$$

Hence, from (59),

$$E_0(x)\,E_2(x)=0.$$

If $E_0(x)=0$, then (21), (24) imply that $E_2(x)\neq 0$, so that $F_4(x)=0$ follows, from (61). This, however, in view of (37) and $\Omega(x)\neq 0$, contradicts our assumption, and therefore

$$E_0(x)\neq 0, \quad E_2(x)=0$$

must hold. Thus, on the one hand, by (21)

$$D_2(x)=E_0(x)$$

and on the other by (61)

$$2D_2(x)^2F_2(x)F_4(x)+D_2(x)^2F_3(x)^2=0,$$

i.e., by (24),

$$2F_2(x)F_4(x)+F_3(x)^2=0.$$

After multiplying by $4\cdot 128$ we have

$$8F_2(x)-128F_4(x)+2(16F_3(x))^2=0,$$

i.e. by (35), (36) and (37)

$$\Omega(x)^2\,\Omega_2(x)+2D_1(x)^2\,\Omega(x)^2=0.$$

Thus, as $\Omega(x)\neq 0$, it follows that $\Omega_2(x)+2D_1(x)^2=0$, and, by (33),

$$7D_1(x)^2=4D_0(x)\,D_2(x).$$

The above results imply that $D_2^\circ = E_0^\circ$, so that (by ($22_2$) and the case $i=0$ of ($20_2$)) the degree estimations

$$D_2^\circ \leqq 2\frac{q_4-1}{P-1}, \quad D_0^\circ \leqq 2\frac{q_3-1}{P-1}, \quad D_1^\circ \leqq \frac{q_3-1}{P-1}+\frac{q_4-1}{P-1}$$

hold. Furthermore, we infer that

$$\Omega_1^\circ \Omega_2^\circ \leqq 2\frac{q_3-1}{P-1}+2\frac{q_4-1}{P-1}$$

from (32) and (33). Thus, by (35)—(37),

$$F_k^\circ \leqq k\frac{q_3-1}{P-1}+\frac{q_4-1}{P-1} \qquad (k = 2, 3, 4).$$

We return to the case $l=2$ of (31):

$$\sum_{k=2}^{3} x^{(3-k)q_3} D_2(x)^{\frac{P+1}{2}-k} F_k(x) \equiv R_2(x) \pmod{x^{2q_3}} \qquad \left(R_2^\circ \leqq 2\frac{q_3-1}{P-1}-1\right).$$

According to the above, the degree of the summand is at most equal to

$$(3-k)q_3+\left(\frac{P+1}{2}-k\right)2\frac{q_4-1}{P-1}+k\left(\frac{q_3-1}{P-1}+\frac{q_4-1}{P-1}\right)$$

$$\left(=(3-k)q_3+k\frac{q_3-1}{P-1}+(P+1-k)\frac{q_4-1}{P-1}\right).$$

It is therefore (by ($3_1$)) certainly less than

$$(4-k)q_3 \qquad (k = 2, 3),$$

i.e. for $k=2$ it is less than $2q_3$, while for $k=3$ it is less than $q_3$. Since, moreover, in this congruence, the first summand:

$$x^{q_3} D_2(x)^{\frac{P-3}{2}} F_2(x)$$

contains the factor $x^{q_3}$ and, by (24) and (35) it is not equal to 0, whereas for the right-hand side $R_2^\circ < q_3$ holds. We therefore have a contradiction, which proves (63).

(37) and (63) imply that

$$F_4(x)=0, \tag{64}$$

so that (61) reduces to

$$E_0(x) D_2(x) F_3(x) = 0.$$

Because of (24), (32), (36) and (57) the second and third factors are different from 0. Therefore

$$E_0(x)=0. \tag{65}$$

So, by (59),

$$E_1(x)=0. \tag{66}$$

($E_2(x)\neq 0$ also follows from (19), (24), (65), (66) which point, however, we shall disregard.)

From (65), (66), it follows from (41) that

$$y^2|\Phi_l(x, y)$$

and thus that

$$x^{2q_3}|\Phi_l(x, x^{q_3}) \qquad (l=1, 2, 3, 4). \tag{67}$$

On the other hand, the degree of the right-hand side of (42) is, according to the estimation deduced before (44), at most equal to

$$(4l+6)\frac{q_3-1}{P-1}-2,$$

and so (by $(3_1)$) certainly less than $2q_3$. Accordingly, the right-hand side of (67) vanishes, and thus from (42):

$$R_l(x)=0 \qquad (l=2, 3, 4). \tag{68}$$

(For $l=1$ this conclusion does not hold!) We return to the congruence (31). By this and (68) we have

$$\sum_{k=2}^{l+1} x^{(l+1-k)q_3} D_2(x)^{\frac{P+1}{2}-k} F_k(x)=0 \pmod{x^{lq_3}} \qquad (l=2, 3, 4). \tag{69}$$

We cannot draw any further conclusions from the cases $l=2, 3$ but the (hitherto disregarded) case $l=4$ leads to the desired contradiction as follows. For $l=4$ we pass over from (69) to the (weaker) congruence mod $x^{2q_3}$. Then only the terms with $k\geqq(l=)4$, i.e. those with $k=4, 5$ are to be retained. As, however, by (64), the term belonging to $k=4$ vanishes, we get simply that

$$D_2(x)^{\frac{P-q}{2}} F_5(x)\equiv 0 \pmod{x^{2q_3}}. \tag{70}$$

The degree of the left-hand side is by $(20_2)$ and (40) at most equal to

$$(P-q)\frac{q_3-1}{P-1}+10\frac{q_3-1}{P-1}\left(=q_3-1+2\frac{q_3-1}{P-1}\right),$$

and thus less than $2q_3$; therefore the left-hand side of (70) vanishes. Hence and from (24)

$$F_5(x)=0$$

follows; thus, by (32), (38) and (57)

$$\Omega_3(x)=0$$

must hold. This equation, together with (63), leads to

$$8D_1(x)^2 = 3\Omega_2(x)-\Omega_3(x) = 0,$$

using (33) and (34); this, however, contradicts (57).

We summarize the result in

THEOREM 15. *Only non-primitive solutions of Problem III belong to a pair $P, \varphi(x)$ with*

$$P \geqq 7, \quad P^3 = q, \quad q^{\downarrow} \geqq 3 \tag{71}$$

*or*

$$P \geqq 19, \quad P^4 \leqq q, \quad \varphi^{\downarrow} \geqq 6. \tag{72}$$

COROLLARY. *Moreover, these solutions are regular.*

(The above proof of Theorem 15 was valid only for the marginal case $\varphi^{\circ} = \dfrac{P-1}{2}$. However, the rest follows, as already pointed out, from Theorem 13.)

Before proving of the corollary we introduce the following definition, which will also come in useful later: A set $\mathscr{M}$ of solutions of Problem III is said to be *closed* (for the successors) if all the successors of its elements likewise belong to $\mathscr{M}$. (In other terms, this means that the property of belonging to $\mathscr{M}$ is hereditary for the formation of successors.)

To prove the corollary, we denote by $\mathscr{M}_1$ the set of those solutions of Problem III which are dealt with in Theorem 15, and which are therefore non-primitive. We have to prove that the elements of $\mathscr{M}_1$, too, are regular. We use the set $\mathscr{M}_2$ of the solutions of Problem III the degree of which is of the form

$$\frac{q-1}{P-1} \qquad (P \underset{1}{|} q, \; P^2 \geqq q).$$

(Of course, $\mathscr{M}_1 \cap \mathscr{M}_2 = \varnothing$.) By Theorem 13, the non-linear elements of $\mathscr{M}_2$ (case $P^2 = q$) are likewise non-primitive. We form the union

$$\mathscr{M}_{12} = \mathscr{M}_1 \cup \mathscr{M}_2$$

and prove that even $\mathscr{M}_{12}$ contains only regular solutions. In view of the definition of regular solutions it suffices to prove that $\mathscr{M}_{12}$ is closed.

For this purpose let us consider an arbitrary $f(x) \in \mathscr{M}_{12}$ and a successor $g(x)$ of it. We must also prove that $g(x) \in \mathscr{M}_{12}$.

$f(x)$ belongs to a pair

$$P, \varphi(x)$$

with $P^2 \leqq q$ or (71) or (72). Furthermore, $g(x)$ belongs to a pair

$$Q, \psi(x)$$

with

$$P \underset{1}{|} Q \underset{1}{|} q, \quad P < Q, \quad \psi(x) = \varphi\left(x^{\frac{Q-1}{P-1}} + \varrho\right) \quad (\varrho \in \mathsf{F}_P).$$

If $Q^2 \geqq q$, then $g(x)$ lies in $\mathscr{M}_2$. If $Q^3 \leqq q$, then, certainly, $P^6 \leqq q$, and thus now $p \geqq 19$, by our assumption; moreover, again by our assumption,

$$\psi^\circ \geqq \frac{q-1}{P-1} > P > 6.$$

Thus we now have $g(x) \in \mathscr{M}_1$. Since we have proved in both cases that $g(x) \in \mathscr{M}_{12}$, the corollary is established.

## § 31. A further part of the marginal case of Problem III without primitive solutions

In what follows we again investigate the marginal case

$$\varphi^\circ = \frac{P-1}{2} \tag{1}$$

of Problem III, doing this here, however, only under the very special assumption

$$\varphi(x) = x^{\frac{P-1}{2}} + \sigma \qquad (\sigma = \pm 1). \tag{2}$$

We shall sometimes call this case the *special* marginal case of Problem III. (According to the Main Lemma, it is precisely the special marginal case of Problem III with $P=p$ that matters if we are concerned with the marginal case of Problem II.) Our aim consists in generalizing Theorem 15 for this special case. By Theorem 13 the cases $P^2 \geqq q$ are known; we thus have to consider only the case

$$P^3 \leqq q. \tag{3}$$

For the case when the equality holds in (3) we can obtain no results that take us beyond Theorem 15. We take

$$P^4 \leqq q \tag{4}$$

and

$$p \geqq 11. \tag{5}$$

We assert that Problem III then has no primitive solutions belonging to the pair $P$, $\varphi(x)$.

(This is already known for $p \geqq 19$ by Theorem 15. Only the cases $p=11$, 13, 17 are new. Accordingly, our assertion could involve only slight progress. This will, however, be of great importance for the marginal case of Problem II.

The proof will be difficult but interesting. It will differ essentially from that of Theorem 15. Although, as stated, it would be sufficient to investigate the cases $p=11$, 13, 17, we admit all the $p$ with (5), since thereby our proof becomes clearer but not longer.) Let us assume the existence of a primitive solution $f(x)$ of Problem III, belonging to $P$, $\varphi(x)$. Several parts of § 23 will be used here without further reference. We disregard (5) for the time being.

The cofactor of $\varphi(x)$ is now, by (2),

$$\bar{\varphi}(x) = x^{\frac{P-1}{2}} - \sigma x. \tag{6}$$

We again use the catalyst $g(x)$ ($\neq 0$) of $f(x)$ which is now defined by

$$f_1(x) g(x) = D(x)^{\frac{P+1}{2}} - \sigma D(x) f_{11}(x)^{\frac{P-1}{2}}, \tag{7}$$

according to (6).

As

$$f_1(x) = f_{10}(x) + x^{q_2} f_{11}(x), \tag{8}$$

(7) becomes, when we multiply out:

$$D(x)^{\frac{P+1}{2}} = \sigma D(x) f_{11}(x)^{\frac{P-1}{2}} + f_{10}(x) g(x) + x^{q_2} f_{11}(x) g(x). \tag{9}$$

Accordingly, we put

$$D(x)^{\frac{P+1}{2}} = D_0(x) + x^{q_2} D_1(x), \tag{10}$$

where

$$D_0(x) = \sigma D(x) f_{11}(x)^{\frac{P-1}{2}} + f_{10}(x) g(x), \quad D_1(x) = f_{11}(x) g(x) (\neq 0). \tag{11}$$

(These $D_0(x)$, $D_1(x)$ could be called the two "constituents" of $D(x)^{\frac{P+1}{2}}$ and of course, have, no connection with the $D_0(x)$, $D_1(x)$, or $D_2(x)$ used in the last section.) The non-vanishing of $D_1(x)$ follows from the corresponding property of $f_1(x)$ and $g(x)$. Later it will turn out that $D_0(x) \neq 0$ is also true.

We endeavour to examine the degree of $D(x)$, $D_0(x)$ and $D_1(x)$. As we already know,

$$f_0^{\circ} \leqq f_1^{\circ} = \frac{q_1 - 1}{P-1}\left(= \frac{Pq_2 - 1}{P-1}\right), \quad f_0^{\circ_P} \leqq f_1^{\circ_P} = 1, \tag{12}$$

$$f_{ij}^{\circ} \leqq f_{11}^{\circ} = \frac{q_2 - 1}{P-1}; \quad f_{ij}^{\circ_P} \leqq f_{11}^{\circ_P} = 1 \qquad (i, j = 0, 1) \tag{13}$$

and

$$D^{\circ} \leqq (2f_{11}^{\circ} =) 2\frac{q_2 - 1}{P-1}. \tag{14}$$

It follows from (14) that

$$\left(D(x)^{\frac{P+1}{2}}\right)^{\circ} \leqq (P+1)\frac{q_2 - 1}{P-1}\left(= q_2 - 1 - 2\frac{q_2-1}{P-1}\right). \tag{15}$$

By $(13_1)$ and (14) we have

$$\left(D(x) f_{11}(x)^{\frac{P-1}{2}}\right)^{\circ} \leqq \frac{P+3}{2}\frac{q_2 - 1}{P-1}. \tag{16}$$

As $P \geqq 3$ and $q_2 > 1$, $(12_1)$ and (16) clearly imply that

$$\left(D(x) f_{11}(x)^{\frac{P-1}{2}}\right)^{\circ} < f_1^{\circ}. \tag{17}$$

So it is clear that in (7) the left-hand side and the first term of the right-hand side are of the same degree:

$$f_1^{\circ} + g^{\circ} = \frac{P+1}{2} D^{\circ}. \tag{18}$$

Thus, by virtue of $(12_1)$ and (18), we have

$$D^{\circ} \geqq 2\frac{q_1 - 1}{P^2 - 1}\left(= 2\frac{Pq_2 - 1}{P^2 - 1}\right). \tag{19}$$

We now show that

$$D_0^{\circ} = \frac{q_2 - 1}{2} + D^{\circ}, \quad D_1^{\circ} = \frac{P+1}{2} D^{\circ} - q_2. \tag{20}$$

To do this, we must first show that

$$f_{10}^{\circ} + g^{\circ} < \frac{P-1}{2} f_{11}^{\circ} + D^{\circ}.$$

From $(13_1)$, it is sufficient to show that

$$g^\circ < \frac{P-3}{2} f_{11}^\circ + D^\circ.$$

According to (18), this inequality is equivalent to

$$\frac{P-1}{2} D^\circ - f_1^\circ < \frac{P-3}{2} f_{11}^\circ.$$

On account of $(13_1)$ and (14), the left-hand side is at most equal to

$$q_2 - 1 - \frac{Pq_2 - 1}{P-1} \left( = -\frac{q_2 + P - 2}{P-1} \right);$$

it is therefore negative and thus the stated inequality is correct. It asserts that on the right-hand side of $(11_1)$ the second term is of smaller degree than the first term and so the latter and the left-hand side are of the same degree. In view of $(13_1)$ this proves $(20_1)$.

From $(11_2)$ and (18) it follows that

$$D_1^\circ = \frac{P+1}{2} D^\circ - f_1^\circ + f_{11}^\circ$$

which gives $(20_2)$, using $(13_1)$ and (14).

From now on we shall also assume (5). We have to deduce a contradiction. We shall do this by showing that the supposed primitive solution is necessarily non-primitive.

We prove first of all

$$\frac{D_1(x)}{D_0(x)} \in \mathsf{F}(x^P). \tag{21}$$

(As we have already mentioned, the numerator and the denominator are not equal to 0.)

Let us assume (21) to be false. Then there exists a divisor

$$P_1 \left| \frac{P}{p} \right. \tag{22}$$

such that

$$\frac{D_1(x)}{D_0(x)} \in \mathsf{F}(x^{P_1}) \setminus \mathsf{F}(x^{pP_1}). \tag{23}$$

This will lead to a contradiction.

(23) implies the existence of polynomials with

$$D_1(x) = A(x)B(x)^{P_1}, \quad D_0(x) = A(x)C(x)^{P_1}, \tag{24}$$

such that $A(x)$ is divisible by the $(P-1)$-th power at most of any prime polynomial from $\mathsf{F}[x]$. (The case $P_1=1$ is no exception, as we may then write $A(x)=1$, $B(x)=D_1(x)$, $l(x)=D_0(x)$.) After substituting into (10) we have

$$D(x)^{\frac{P+1}{2}} = A(x)\left(x^{q_2}B(x)^{P_1}+C(x)^{P_1}\right).$$

So, from the above properties of $A(x)$ we infer the divisibility

$$D(x)^{\frac{P+1}{2}-(P_1-1)} \Big| x^{q_2}B(x)^{P_1}+C(x)^{P_1}.$$

The exponent of the left-hand side is equal to

$$\frac{P+3}{2} - P_1 \left( \geqq \frac{pP_1+3}{2} - P_1 > \frac{9P_1}{2} \right)$$

[where (5) has been used in the brackets]. Thus

$$D(x)^{P_1 l} | x^{q_2}B(x)^{P_1}+C(x)^{P_1}$$

holds for the natural number

$$l = \left[\frac{P+3}{2P_1}\right] - 1. \tag{25}$$

After extraction of the $P_1$-th root we have

$$D(x)^l | x^{P_1^{-1}q_2}B(x)+C(x).$$

By virtue of (4), i.e. $P^2|q_2$, the exponent on the right-hand side is divisible by $p$ (it is in fact divisible by $pP$, but the weaker statement will suffice). Thus by differentiation we get

$$D(x)^{l-1} | x^{P_1^{-1}q_2}B'(x)+C'(x).$$

From these two results

$$D(x)^{l-1} \left| \begin{vmatrix} B(x) & C(x) \\ B'(x) & C'(x) \end{vmatrix} \right. \tag{26}$$

follows by Lemma I.

If the right-hand side vanishes, it follows by quadrature that

$$\frac{B(x)}{C(x)} \in \mathsf{F}(x^p)$$

holds. On the other hand, by (24)

$$\frac{D_1(x)}{D_0(x)} = \left(\frac{B(x)}{C(x)}\right)^{P_1}.$$

From these two results it follows that

$$\frac{D_1(x)}{D_0(x)} \in \mathsf{F}(x^{pP_1}).$$

As this is false, by (23), it follows that the right-hand side of (26) is not equal to 0.

In particular, it follows from this that at least one of $B(x)$ and $C(x)$ is not constant, and thus, by (26), that

$$(l-1)D^\circ \leqq B^\circ + C^\circ - 1,$$

by degree comparison. After multiplying by $P_1$ we have

$$(l-1)P_1D^\circ \leqq P_1B^\circ + P_1C^\circ - P_1$$

and thus

$$(l-1)P_1D^\circ \leqq D_1^\circ + D_0^\circ - P_1$$

follows from (24). By substituting (20) we get

$$(l-1)P_1D^\circ \leqq \frac{P+3}{2}D^\circ - \frac{q_2+1}{2} - P_1,$$

i.e.

$$\frac{q_2+1}{2} + P_1 \leqq \left(\frac{P+3}{2} - (l-1)P_1\right)D^\circ.$$

By (14), we obtain

$$\frac{q_2+1}{2} + P_1 \leqq (P+3-2(l-1)P_1)\frac{q_2-1}{P-1}. \tag{27}$$

We shall deduce the required contradiction from this inequality.

First let $P_1 > 1$. Then by (25),

$$l \geqq \frac{P-P_1}{2P_1} - 1 = \frac{P-3P_1}{2P_1}.$$

After substitution into (27) we have

$$\frac{q_2+1}{2} + P_1 \leqq (P+3-(P-5P_1))\frac{q_2-1}{P-1} = (5P_1+3)\frac{q_2-1}{P-1}.$$

Multiplying by $2(P-1)$ we obtain

$$(q_2+1+2P_1)(P-1) \leqq (10P_1+6)(q_2-1),$$

i.e.

$$(P-10P_1-7)q_2 \leqq -P(1+2P_1)-8P_1-5.$$

The left-hand side is positive, from $P \geqq pP_1$, $P_1 \geqq p$ and (5). This is a contradiction, since the right-hand side is negative.

In the remaining case $P_1=1$, by virtue of (25),

$$\Delta = \frac{P+1}{2}$$

holds, so that (27) becomes

$$\frac{q_2+3}{2} \leqq 4\,\frac{q_2-1}{P-1}.$$

It follows that

$$(q_2+3)(P-1) \leqq 8q_2-8,$$

i.e. that

$$(P-9)q_2 \leqq -3P-5.$$

This is, by virtue of $P \geqq p$ and (5), a contradiction similar to the last one, which proves (21).

We now eliminate the catalyst $g(x)$ from the system of equations (11). This gives us

$$\begin{vmatrix} D_0(x)-\sigma D(x)f_{11}(x)^{\frac{P-1}{2}} & D_1(x) \\ f_{10}(x) & f_{11}(x) \end{vmatrix} = 0.$$

Hence and from $(13_1)$

$$\left(D_0(x)-D(x)f_{11}(x)^{\frac{P-1}{2}}\right)^{\circ} \leqq D_1^{\circ}. \tag{28}$$

We now write

$$U_0(x) = f_{10}(x)g(x), \quad U_1(x) = -D_0(x), \tag{29}$$

which will put the succeeding formulae into more manageable shape. From (10) and $(11_1)$ we infer that

$$D_0(x) = \sigma D(x)f_{11}(x)^{\frac{P-1}{2}}+U_0(x), \quad x^{q_2}D_1(x) = D(x)^{\frac{P+1}{2}}+U_1(x). \tag{30}$$

We shall show that

$$U_0^{\circ} \leqq \frac{P+1}{2}D^{\circ}-q_2, \quad U_1^{\circ} = \frac{q_2-1}{2}+D^{\circ}. \tag{31}$$

From ($11_2$), ($13_1$) and ($29_1$) $U_0^\circ \leqq D_1^\circ$ follows. Hence and from ($20_2$) ($31_1$) follows. By ($20_1$) and ($29_2$) it can be seen that ($31_2$) also holds.

We define $\mathfrak{p}(x)$ by

$$\mathfrak{p}(x)^P = \frac{D(x)^{\frac{P+1}{2}} + U_1(x)}{D(x)f_{11}(x)^{\frac{P-1}{2}} + \sigma U_0(x)} \qquad \left(\mathfrak{p}(x) \in \mathsf{F}\left(\left(\frac{1}{x}\right)\right)\right). \tag{32}$$

As the right-hand side is by (30) and $\sigma = \pm 1$ equal to

$$\sigma \frac{x^{q_2} D_1(x)}{D_0(x)}$$

and $P|q_2$, it follows from (21) that $\mathfrak{p}(x)$ exists. It is uniquely determined and it clearly has positive degree $\mathfrak{p}^\circ$; moreover

$$P\mathfrak{p}^\circ = q_2 + D_1^\circ - D_0^\circ.$$

Hence and from (20)

$$\mathfrak{p} = \frac{P-1}{2P}\left(D^\circ - \frac{q_2 - 1}{P-1}\right) (>0) \tag{33}$$

where (19), too, has been applied. It follows that

$$\frac{P-1}{2} \Bigm| \mathfrak{p}^\circ. \tag{34}$$

((33) also implies that $D^\circ \equiv 1 \pmod{P}$, which can alternatively be obtained directly from (20) and (21), but which will not be used in what is to follow.) We put

$$r = \frac{2\mathfrak{p}^\circ}{P-1}. \tag{35}$$

By (34), this $r$ is a natural number.

Our next object is to obtain certain information concerning $D(x)$ from (32), (33). We write (32) in the form

$$D(x)^{\frac{P+1}{2}} = \left(D(x)f_{11}^{\frac{P-1}{2}} + \sigma U_0(x)\right)\mathfrak{p}(x)^P - U_1(x).$$

(Here and in what follows, where we are dealing with equations of the form $X = y+z$, $y$ and $z$ should be thought of as the "main term" and the "remainder term" of $X$, respectively.) It follows that

$$\left(\frac{D(x)}{f_{11}(x)}\right)^{\frac{P-1}{2}} = \mathfrak{p}(x)^P + \frac{\sigma U_0(x)\mathfrak{p}(x)^P - U_1(x)}{D(x)f_{11}(x)^{\frac{P-1}{2}}}$$

Just for a moment let us write $k$ for the degree of the remainder term appearing here and let us estimate it from above. We have

$$k \leqq -D^{\circ} - \frac{P-1}{2} f_{11}^{\circ} + \max (U_0^{\circ} + P\mathfrak{p}^{\circ}, U_1^{\circ}).$$

Thus, by virtue of ($13_1$), (31) and (33),

$$k \leqq -D^{\circ} - \frac{q_2-1}{2} + \max \left( \frac{P+1}{2} D^{\circ} - q_2 + \frac{P-1}{2} \left( D^{\circ} - \frac{q_2-1}{P-1} \right), \frac{q_2-1}{2} + D^{\circ} \right),$$

i.e. after a few transformations

$$k \leqq \max ((P-1) D^{\circ} - 2q_2 + 1, 0).$$

Hence, by (14),

$$k \leqq \max (2(q_2-1) - 2q_2 + 1, 0) = \max (-1, 0) = 0.$$

Thus

$$\left( \frac{D(x)}{f_{11}(x)} \right)^{\frac{P-1}{2}} = \mathfrak{p}(x)^P + \mathfrak{q}(x) \qquad \left( \mathfrak{q}(x) \in \mathsf{F}\left(\left(\frac{1}{x}\right)\right), \ \mathfrak{q}^{\circ} \leqq 0 \right)$$

follows from the preceding equation with a suitable $\mathfrak{q}(x)$.

We transform this equation into

$$\left( \frac{D(x)}{x^{Pr} f_{11}(x)} \right)^{\frac{P-1}{2}} = \left( \frac{\mathfrak{p}(x)}{x^{\frac{P-1}{2}r}} \right)^P + \frac{\mathfrak{p}(x)}{x^{P\frac{P-1}{2}r}}. \tag{36}$$

The degrees of the main and remainder terms are not greater than zero and

$$-P\frac{P-1}{2} r,$$

respectively. Thus from (36) the existence of an equation

$$\frac{D(x)}{x^{Pr} f_{11}(x)} = \mathfrak{u}(x)^P + \frac{\mathfrak{v}(x)}{x^{P\frac{P-1}{2}r}},$$

i.e.

$$D(x) = x^{Pr} f_{11}(x) \mathfrak{u}(x)^P + \frac{f_{11}(x) \mathfrak{v}(x)}{x^{P\frac{P-3}{2}r}}$$

$$\left( \mathfrak{u}(x), \mathfrak{v}(x) \in \mathsf{F}\left(\left(\frac{1}{x}\right)\right); \ \mathfrak{u}^{\circ} = 0, \ \mathfrak{v}^{\circ} \leqq 0 \right)$$

can clearly be inferred with suitable $\mathfrak{u}(x)$, $\mathfrak{v}(x)$. After changing the notation for the remainder term this equation can be written simply as follows:

$$D(x) = x^{Pr} f_{11}(x)\mathfrak{u}(x) + \mathfrak{v}(x) \qquad \left(\mathfrak{u}(x),\, \mathfrak{v}(x) \in \mathsf{F}\left(\left(\frac{1}{x}\right)\right);\ \mathfrak{u}^\circ = 0\right), \tag{37}$$

where the degree estimation

$$\mathfrak{v}^\circ \leqq -P\frac{P-3}{2}r + f_{11}^\circ$$

holds; we shall, however, immediately transform this estimation. From ($13_1$), (33) and (35) we obtain

$$\mathfrak{v}^\circ \leqq -P\frac{P-3}{P-1}\mathfrak{p}^\circ + \frac{q_2-1}{P-1} = -\frac{P-3}{2}\left(D^\circ - \frac{q_2-1}{P-1}\right) + \frac{q_2-1}{P-1},$$

i.e.

$$\mathfrak{v}^\circ \leqq -\frac{P-3}{2}D^\circ + \frac{q_2-1}{2}. \tag{38}$$

Applying the lower estimation (19) we deduce from (38)

$$\mathfrak{v}^\circ \leqq -(P-3)\frac{Pq_2-1}{P^2-1} + \frac{q_2-1}{2},$$

i.e.

$$\mathfrak{v}^\circ \leqq \frac{-(P^2-6P+1)q_2-(P^2-2P+5)}{2(P^2-1)}. \tag{39}$$

We shall draw two conclusions from the result given in (37) and (39). It will be necessary to use not only the one-step and two-step constituents $\big(f_i(x)$ and $f_{ij}(x)\big)$ of $f(x)$, but also its *three-step constituents* $f_{ijk}(x)$ $(i, j, k=0, 1)$ which, using the fact that

$$f(x) \underset{1}{\in} \mathsf{F}[x], \quad f^\circ = \frac{q-1}{P-1}, \quad f^{\circ_P} = 1, \tag{40}$$

we define by the "decomposition formula"

$$f(x) = \sum_{i,j,k=0}^{1} x^{iq_1+jq_2+kq_3} f_{ijk}(x) \in \mathsf{F}[x], \quad f_{ijk}^\circ \leqq \frac{q_3-1}{P-1} \tag{41}$$

where—as above—it is understood that $q_3 = P^{-3}q$. Uniqueness would, of course, be assured here already by the weaker requirement $f_{ijk}^\circ < q_3$. (The reader should observe the analogy with the conditions for Proposition 25.)

We show that from

$$x^q - x \mid f(x)^P - f(x) \tag{42}$$

we have the decomposition

$$f(x) = \sum_{i,j,k=0}^{1} x^{iP^2+jP+k} f_{ijk}(x)^{P^3} \tag{43}$$

*dual* to (41).

To prove this, we write the right-hand side of (43) as $u(x)$. Then we have to prove

$$f(x) = u(x).$$

We clearly have that

$$u^\circ \leqq P^2+P+1+P^3\frac{q_3-1}{P-1} = \frac{q-1}{P-1}.$$

Therefore, as $f(x)$, $u(x)$ are of degree $<q$, it suffices to prove that

$$f(\xi) = u(\xi)$$

for all $\xi \in \mathsf{F}$. As, however, from (42) $f(x)$ is an $(\mathsf{F}, \mathsf{F}_P)$-polynomial, all the $f(\xi)$ lie in $\mathsf{F}_P$, so that $f(\xi) = f(\xi)^P$ always holds. Thus, by (41),

$$f(\xi) = f(\xi)^{P^3} = \sum_{i,j,k=0}^{1} \xi^{iqP^2+jqP+kq} f_{ijk}(\xi)^{P^3}.$$

In the exponents of $\xi$ the factors $q$ can be cancelled, as $\xi^q = \xi$, whence we obtain $f(\xi) = u(\xi)$. This proves (43).

We note that the formulae

$$f(x) = \sum_{i=0}^{1} x^{iq_1} f_i(x) = \sum_{i,j=0}^{1} x^{iq_1+jq_2} f_{ij}(x) \quad \left(f_i^\circ \leqq \frac{q_1-1}{P-1},\ f_{ij} \leqq \frac{q_2-1}{P-1}\right) \tag{44}$$

and

$$f(x) = \sum_{i=0}^{1} x^i f_i(x)^P = \sum_{i,j=0}^{1} x^{iP+j} f_{ij}(x)^{P^2} \tag{45}$$

analogous to (41) and (43), are valid and can be deduced in a similar way. (These formulae are not new but only summarize ones that we have already used repeatedly.)

From the comparison of (41), (43), (44), (45) with one another the further (dual) formulae

$$f_i(x) = f_{i0}(x) + x^{q_2} f_{i1}(x), \quad f_{ij}(x) = f_{ij0}(x) + x^{q_3} f_{ij1}(x), \tag{46}$$

$$f_i(x) = f_{0i}(x)^P + x f_{1i}(x)^P, \quad f_{ij}(x) = f_{0ij}(x)^P + x f_{1ij}(x)^P \tag{47}$$

follow (the first half of which was also known already).

It will be of importance that the dual formulae

$$D(x) = \begin{vmatrix} f_{000}(x)+x^{q_3}f_{001}(x) & f_{010}(x)+x^{q_3}f_{011}(x) \\ f_{100}(x)+x^{q_3}f_{101}(x) & f_{110}(x)+x^{q_3}f_{111}(x) \end{vmatrix}, \tag{48}$$

$$D(x) = \begin{vmatrix} f_{000}(x)^P+xf_{100}(x)^P & f_{001}(x)^P+xf_{101}(x)^P \\ f_{010}(x)^P+xf_{110}(x)^P & f_{011}(x)^P+xf_{111}(x)^P \end{vmatrix} \tag{49}$$

come from

$$D(x) = \begin{vmatrix} f_{00}(x) & f_{01}(x) \\ f_{10}(x) & f_{11}(x) \end{vmatrix} \tag{50}$$

by substituting from $(46_2)$ and $(47_2)$.

Now in order to be able to obtain from (37) and (39) the desired results, we use the fact that since $f_{11}^{\circ P}=1$ the exponent set of the main term of (37) contains only elements congruent mod $P$ to 0 or 1, while that of the remainder term contains, because of (5) and (39), only negative elements. By (50) this verifies the validity of

$$\begin{vmatrix} f_{100}(x)^P & f_{101}(x)^P \\ f_{110}(x)^P & f_{111}(x)^P \end{vmatrix} = 0,$$

i.e. (after extraction of the $P$-th root and a reflection on the diagonal)

$$\begin{vmatrix} f_{100}(x) & f_{110}(x) \\ f_{101}(x) & f_{111}(x) \end{vmatrix} = 0. \tag{51}$$

This is the "first" conclusion we wished to draw from (37), (39).

To be able to deduce the "second" conclusion, we must observe that, by (51), (50) reduces to

$$D(x) = F_0(x)^P+xF_1(x)^P, \tag{52}$$

where

$$F_0(x) = \begin{vmatrix} f_{000}(x) & f_{001}(x) \\ f_{010}(x) & f_{011}(x) \end{vmatrix}, \quad F_1(x) = \begin{vmatrix} f_{000}(x) & f_{101}(x) \\ f_{010}(x) & f_{111}(x) \end{vmatrix} + \begin{vmatrix} f_{100}(x) & f_{001}(x) \\ f_{110}(x) & f_{011}(x) \end{vmatrix}. \tag{53}$$

Substituting (52) and the case $i=j=1$ of $(47_2)$, equation (37) runs as follows:

$$F_0(x)^P+xF_1(x)^P = x^{Pr}\big(f_{011}(x)^P+xf_{111}(x)^P\big)\mathfrak{u}(x)^P+\mathfrak{v}(x).$$

We see (by component comparison) that $\mathfrak{v}(x)$ must necessarily be of the form

$$\mathfrak{v}(x) = \mathfrak{v}_0(x)^P+x\mathfrak{v}_1(x)^P \qquad \left(\mathfrak{v}_i(x)\in \mathsf{F}\left(\left(\frac{1}{x}\right)\right),\ \mathfrak{v}_i^{\circ} \leqq \frac{\mathfrak{v}^{\circ}}{P}\right) \tag{54}$$

and that, moreover, (after going over to reduced components) the equations

$$F_i(x) = x^r f_{i11}(x)\mathfrak{u}(x)+\mathfrak{v}_i(x) \qquad (i = 0, 1)$$

hold. By eliminating $\mathfrak{u}(x)$ we get

$$\begin{vmatrix} F_0(x) & f_{011}(x) \\ F_1(x) & f_{111}(x) \end{vmatrix} = \begin{vmatrix} \mathfrak{v}_0(x) & f_{011}(x) \\ \mathfrak{v}_1(x) & f_{111}(x) \end{vmatrix}. \tag{55}$$

We shall temporarily write $l$ for the degree of the right-hand side and estimate it from above. We see that (cf. (41) and (54))

$$l \leqq \frac{q_3-1}{P-1} + \frac{\mathfrak{v}^\circ}{P}.$$

Hence and from (39) (as $q_2 = Pq_3$) it follows that

$$l < \frac{q_3}{P-1} - \frac{(P^2-6P+1)q_3}{2(P^2-1)} = -\frac{P^2-8P-1}{2(P^2-1)}q_3.$$

Thus $l<0$, by (5). As, on the other hand, the left-hand side of (55) is a polynomial, it must vanish. Thus we have

$$f_{111}(x)F_0(x) = f_{011}(x)F_1(x).$$

After substituting from (53) and multiplying out, we get

$$f_{111}f_{000}f_{011} - f_{111}f_{001}f_{010} = f_{011}f_{010}f_{111} - f_{011}f_{101}f_{010} + f_{011}^2 f_{100} - f_{011}f_{001}f_{110}$$

using the abbreviation $f_{ijk}(x) = f_{ijk}$.

On both sides the first terms are equal to one another, consequently they can be cancelled. A rearrangement of the terms gives

$$f_{011}^2 f_{100} - f_{011}f_{101}f_{010} - f_{011}f_{001}f_{110} + f_{111}f_{001}f_{010} = 0.$$

Multiplying by $f_{111}$, we get

$$f_{011}^2 f_{100}f_{111} - f_{011}f_{101}f_{010}f_{111} - f_{011}f_{001}f_{110}f_{111} + f_{111}^2 f_{001}f_{010} = 0.$$

Taking into consideration in the first term that, by (51)

$$f_{100}f_{111} = f_{101}f_{110},$$

we obtain

$$(f_{011}f_{101} - f_{001}f_{111})(f_{011}f_{110} - f_{010}f_{111}) = 0,$$

i.e.

$$\begin{vmatrix} f_{001}(x) & f_{011}(x) \\ f_{101}(x) & f_{111}(x) \end{vmatrix} \begin{vmatrix} f_{010}(x) & f_{110}(x) \\ f_{011}(x) & f_{111}(x) \end{vmatrix} = 0. \tag{56}$$

This equation is the second conclusion to be drawn from (37) and (39), but we shall show very simply by another method that the second factor must vanish.

If the assertion is false then the first factor in (56) will vanish. This means that on the right-hand side of (49) (after evaluating the determinant and multiplying out) the cofactor of $x^{2q_3}$ vanishes. Hence (cf. (41)) the degree estimation

$$D^\circ \leqq q_3 + 2\frac{q_3-1}{P-1}$$

follows. Comparison with (19) gives (as $q_2 = Pq_3$)

$$2\frac{P^2 q_3 - 1}{P^2-1} \leqq q_3 + 2\frac{q_3-1}{P-1},$$

i.e., after multiplying by $P^2-1$

$$2P^2 q_3 - 2 \leqq P^2 q_3 - q_3 + 2Pq_3 + 2q_3 - 2P - 2,$$

so that

$$(P^2-2P-1)q_3 \leqq -2P.$$

This contradiction proves that in (56) the second factor vanishes.

This result, together with (51), states that the two second order subdeterminants that include the third column of the matrix

$$\begin{pmatrix} f_{100}(x) & f_{010}(x) & f_{110}(x) \\ f_{101}(x) & f_{011}(x) & f_{111}(x) \end{pmatrix} \tag{57}$$

vanish. It follows from this that all three second-order subdeterminants of this matrix vanish, unless the third column vanishes. As this does not occur, since $f_{111}(x) \neq 0$, we have:

$$\begin{vmatrix} f_{100}(x) & f_{010}(x) \\ f_{101}(x) & f_{011}(x) \end{vmatrix} = 0, \quad \begin{vmatrix} f_{100}(x) & f_{110}(x) \\ f_{101}(x) & f_{111}(x) \end{vmatrix} = 0, \quad \begin{vmatrix} f_{010}(x) & f_{110}(x) \\ f_{011}(x) & f_{111}(x) \end{vmatrix} = 0. \tag{58}$$

However strict these conditions may be, in order to obtain a contradiction, we still need more information, which we shall take from other sources. On the right-hand side of ($53_2$) the secondary diagonals may be interchanged. Thus

$$F_1(y) = \begin{vmatrix} f_{000}(x) & f_{001}(x) \\ f_{110}(x) & f_{111}(x) \end{vmatrix} + \begin{vmatrix} f_{100}(x) & f_{101}(x) \\ f_{010}(x) & f_{001}(x) \end{vmatrix}$$

arises. The second determinant on the right-hand side is equal to 0 by ($58_1$); therefore, together with ($53_1$), we have

$$F_0(x) = \begin{vmatrix} f_{000}(x) & f_{001}(x) \\ f_{010}(x) & f_{011}(x) \end{vmatrix}, \quad F_1(x) = \begin{vmatrix} f_{000}(x) & f_{001}(x) \\ f_{110}(x) & f_{111}(x) \end{vmatrix}. \tag{59}$$

As the two first rows are equal, we see from (52) that

$$D(x) = \begin{vmatrix} f_{000}(x)^P & f_{001}(x)^P \\ f_{010}(x)^P + xf_{111}(x)^P & f_{011}(x)^P + xf_{111}(x)^P \end{vmatrix}.$$

The second row consists according to the cases $i=1$ and $j=0, 1$ of $(47_2)$ just of the two-step constituents $f_{10}(x)$, $f_{11}(x)$. Thus

$$D(x) = f_{000}(x)^P f_{11}(x) - f_{001}(x)^P f_{10}(x). \tag{60}$$

A major change is achieved by considering the divisibility

$$f_1(x) \Big| \prod_{\varphi(\varrho)=} (f_0(x) - \varrho_1 f_1) \tag{61}$$

by Proposition 21. (For the cofactor $\bar{\varphi}(x)$ of $\varphi(x)$ (6) now holds, but this will be disregarded here.) Here we treat (61) somewhat differently, as in § 23. (Indeed, in what follows, we shall be able to draw a large number of conclusions from simultaneous use of the two- and three-step constituents of $f(x)$.)

Combination of the formulae $(46_1)$ and $(46_2)$ yields

$$f_i(x) (= f_{i0}(x) + x^{q_2} f_{i1}(x)) = f_{i0}(x) + x^{q_2} f_{i10}(x) + x^{q_2+q_3} f_{i11}(x) \qquad (i = 0, 1).$$

We again return to Lemma I which we shall use with $k = q_2 + q_3$. Thus from these formulae we get the divisibility

$$(f_0(x) - \varrho, f_1(x)) \Big| \begin{vmatrix} f_{00}(x) + x^{q_2} f_{010}(x) - \varrho & f_{011}(x) \\ f_{10}(x) + x^{q_2} f_{110}(x) & f_{111}(x) \end{vmatrix}.$$

Now $(58_3)$ states that (after explication) the cofactor of $x^{q_3}$ vanishes here whence it follows that

$$(f_0(x) - \varrho, f_1(x)) \Big| \begin{vmatrix} f_{00}(x) - \varrho & f_{011}(x) \\ f_{10}(x) & f_{111}(x) \end{vmatrix}.$$

(Although all this holds for every $\varrho \in \mathsf{F}$, we use only the $\varrho$ satisfying $\bar{\varphi}(\varrho) = 0$.) Thus, from (61)

$$f_1(x) \Big| \prod_{\bar{\varphi}(\varrho)=0} \begin{vmatrix} f_{00} - \varrho & f_{011}(x) \\ f_{10}(x) & f_{111}(x) \end{vmatrix}. \tag{62}$$

The degree of the right-hand side is (cf. (41) and (44)) at most equal to

$$\frac{P+1}{2} \left( \frac{q_2 - 1}{P-1} + \frac{q_3 - 1}{P-1} \right).$$

That of the left-hand side is by (12)

$$\frac{q_1-1}{P-1}.$$

Since this number is greater than the former one because of $q_1=Pq_2=P^2q_3$, the right-hand side of (62) vanishes. This means that there exists a $\varrho\in \mathsf{F}_P$ such that $\bar{\varphi}(\varrho)=0$ and

$$\begin{vmatrix} f_{00}(x)-\varrho & f_{011}(x) \\ f_{10}(x) & f_{111}(x) \end{vmatrix} = 0. \tag{63}$$

We shall use this (uniquely determined) value of $\varrho$ from now on.

Fortunately (63) can be written in a similar form in which only the three-step constituents $f_{ijk}(x)$ appear. For, by repeated application of ($46_2$), equation (63) gives:

$$\begin{vmatrix} f_{000}(x)+x^{q_3}f_{001}(x)-\varrho & f_{011}(x) \\ f_{100}(x)+x^{q_3}f_{101}(x) & f_{111}(x) \end{vmatrix} = 0.$$

Thus we have

$$\begin{vmatrix} f_{000}(x)-\varrho & f_{011}(x) \\ f_{100}(x) & f_{111}(x) \end{vmatrix} + x^{q_3} \begin{vmatrix} f_{001}(x) & f_{011}(x) \\ f_{101}(x) & f_{111}(x) \end{vmatrix} = 0.$$

The first term is [cf. (41)] of degree

$$\leqq 2\,\frac{q_3-1}{P-1}\,(< q_3),$$

so that (63) finally becomes

$$\begin{vmatrix} f_{000}(x)-\varrho & f_{011}(x) \\ f_{100}(x) & f_{111}(x) \end{vmatrix} = 0, \qquad \begin{vmatrix} f_{001}(x) & f_{011}(x) \\ f_{101}(x) & f_{111}(x) \end{vmatrix} = 0. \tag{64}$$

From ($58_2$), ($58_3$), ($64_1$), ($64_2$) we obtain (as $f_{111}(x)\neq 0$)

$$f_{100}(x) = \frac{f_{110}(x)f_{101}(x)}{f_{111}(x)}, \quad f_{010}(x) = \frac{f_{110}(x)f_{011}(x)}{f_{111}(x)},$$

$$f_{000}(x) = \frac{f_{011}(x)f_{100}(x)}{f_{111}(x)}+\varrho, \quad f_{001}(x) = \frac{f_{011}(x)f_{101}(x)}{f_{111}(x)}.$$

Substituting the first of these four formulae into the third one, the latter becomes

$$f_{000}(x) = \frac{f_{011}(x)f_{110}(x)f_{101}(x)}{f_{111}(x)^2}+\varrho.$$

Hence, and from the other three equations, we obtain, after removing the denominator, the four formulae

$$\begin{cases} f_{100}(x)f_{111}(x)=f_{110}(x)f_{101}(x), & f_{010}(x)f_{111}(x)=f_{110}(x)f_{011}(x), \\ f_{001}(x)f_{111}(x)=f_{011}(x)f_{101}(x), & f_{000}(x)f_{111}(x)^2=f_{011}(x)f_{110}(x)f_{101}(x)+\varrho f_{111}(x)^2 \end{cases} \tag{65}$$

(which we have written in a convenient order).

We next multiply equations (59) by $f_{111}(x)^2$ as follows:

$$F_0(x)f_{111}(x)^2 = \begin{vmatrix} f_{000}(x)f_{111}(x)^2 & f_{001}(x)f_{111}(x) \\ f_{010}(x)f_{111}(x) & f_{011}(x) \end{vmatrix},$$

$$F_1(x)f_{111}(x)^2 = \begin{vmatrix} f_{000}(x)f_{111}(x)^2 & f_{001}(x)f_{111}(x) \\ f_{110}(x)f_{111}(x) & f_{111}(x) \end{vmatrix}.$$

Substitution of (65) gives the equations

$$F_0(x)f_{111}(x)^2 = \begin{vmatrix} f_{011}(x)f_{110}(x)f_{101}(x)+\varrho f_{111}(x)^2 & f_{011}(x)f_{101}(x) \\ f_{110}(x)f_{011}(x) & f_{011}(x) \end{vmatrix}$$

and

$$F_1(x)f_{111}(x)^2 = \begin{vmatrix} f_{011}(x)f_{110}(x)f_{101}(x)+\varrho f_{111}(x)^2 & f_{011}(x)f_{101}(x) \\ f_{110}(x)f_{111}(x) & f_{111}(x) \end{vmatrix}.$$

In both determinants we multiply the second column by $f_{110}(x)$ and subtract the product from the first column. Finally, after dividing by $f_{111}(x)^2$, we obtain

$$F_0(x)=\varrho f_{011}(x), \quad F_1(x)=\varrho f_{111}(x).$$

Substituting this into (52), we get

$$D(x) = \varrho^P(f_{011}(x)^P+xf_{111}(x)^P)$$

thus by $(47_2)$

$$D(x)-\varrho f_{11}(x) = 0, \tag{66}$$

where we have used the fact that $\varrho^P$ is equal to $\varrho$, because $\varrho \in \mathsf{F}_P$. As the left-hand side is one of the codeterminants $D_P(x)$ of $f(x)$, by Theorem 12, (66) means that the given (primitive) solution $f(x)$ of Problem III is non-primitive. This contradiction proves

THEOREM 16. (Completion of Theorem 15.) *To a pair $P$, $\varphi(x)$ such that*

$$P \geqq 11, \quad P^4 \leqq q, \quad x^{\frac{P-1}{2}}+\varrho | \varphi(x) \qquad (\sigma=+1) \tag{67}$$

*there belong only non-primitive solutions of Problem III.*

COROLLARY. *Moreover, these solutions are regular.*

(The proof of Theorem 16 was valid only for the case (2), but the rest follows from Theorem 13.)

To prove the corollary, we write as $\mathcal{M}_0$ the set of those solutions of Problem III given by Theorem 16, which are therefore non-primitive. It remains to be proved that the elements of $\mathcal{M}_0$ are also regular. The proof is a slight extension of that given for the corollary of Theorem 15. We use the notations of that proof and form the union

$$\mathcal{M}_{012} = \mathcal{M}_0 \cup \mathcal{M}_{12} (= \mathcal{M}_0 \cup \mathcal{M}_1 \cup \mathcal{M}_2),$$

the elements of which are therefore linear or non-primitive. It is sufficient to prove that $\mathcal{M}_{012}$ is closed.

As compared with earlier proof, we now have to take (not only the $f(x) \in \mathcal{M}_{12}$ but also) the $f(x) \in \mathcal{M}_0$ into consideration. It remains only to prove that $g(x)$ lies in $\mathcal{M}_{012}$. In the case $Q^2 \geqq q$ we have $g(x) \in \mathcal{M}_2$, like earlier. In the other case $Q^3 = q$ or $Q^4 \leqq q$ holds. Since, on the other hand,

$$Q \geqq P^2 \geqq 11^2 > 19$$

and

$$\psi^{\downarrow} \geqq \frac{Q-1}{P-1} > P \geqq 11,$$

we now have $g(x) \in \mathcal{M}_1$.

Thus the corollary is proved.

CHAPTER V

# THE SOLUTION OF PROBLEM II IN ALMOST ALL CASES

As a result of the preceding two chapters we shall here obtain all the solutions of Problem II. Only the characteristics $p=3, 5, 7$ remain as an exception, inasmuch as in these cases (if, in addition, the degree $n$ of $\mathsf{F}$ is at least 4) other solutions may exist, apart from those to be given in this chapter. The degenerate solutions of Problem II will be considered in the Appendix.

## § 32. The regular solutions of Problem II

Let $\chi$ denote the (single) quadratic character of the prime field $\mathsf{F}_p$. Since, as usual, its values 0, 1, $-1$ are also treated as elements of $\mathsf{F}_p$, $\chi$ can be defined by

$$\chi(a) = a^{\frac{p-1}{2}} \qquad (a \in \mathsf{F}_p).$$

$\left[\chi(a) \text{ can also be identified with the Legendre symbol } \left(\frac{a}{p}\right).\right]$

According to the Main Lemma, apart from $x^q - x$ all solutions of Problem II belong to the marginal case and are given by

$$f(x) = \frac{x^q - x}{c(x)^{\frac{p-1}{2}} + \sigma}\left(c(x)^{\frac{p-1}{2}} - \sigma\tau\right) \qquad (\sigma = \pm 1;\ \tau = 0, 1), \tag{1}$$

where $c(x)$ runs through the solutions of Problem III belonging to the pair

$$(P=)p, \quad (\varphi(x)=)x^{\frac{p-1}{2}} + \sigma, \tag{2}$$

only those $f(x)$ being retained for which the second factor of the right-hand side of (1) (i.e. also $f(x)$) is fully reducible. If this is the case for a pair $\tau$, $c(x)$, then $c(x)$ is said to be *friendly* towards this $\tau$. In this definition there is no need to mention $\sigma$, as: For any arbitrary $\sigma = \pm 1$ every

$$c(x) = (x + \varkappa)^{\frac{q-1}{p-1}} \qquad (\varkappa \in \mathsf{F}) \tag{3}$$

is friendly towards both values $\tau = 0, 1$ and $\sigma$ is uniquely determined by any other $c(x)$.

Before proceeding to prove this, we point out that all the $c(x)$ (not only the friendly ones) are defined by

$$c(x) \underset{1}{\in} \mathsf{F}[x], \quad c^{\circ} = \frac{q-1}{p-1}, \quad c^{\circ_p} = 1, \quad x^q - x | c(x)^p - c(x) \tag{4}$$

and

$$c(x)^{\frac{p-1}{2}} + \sigma | x^q - x \tag{5}$$

(where (4) states that $c(x)$ is a stem polynomial).

Now if (3) holds, then (4) and (5) are trivially satisfied for all $\sigma = \pm 1$ and $\tau = 0$, 1, and (1) is obviously fully reducible. This proves the first half of the assertion.

In order to prove the second half, we assume that (5) is satisfied by both values $\sigma = \pm 1$ for a $c(x)$. It is sufficient to prove that (3) then also holds. Since the polynomials given by (5) for $\sigma = \pm 1$ are prime to one another, the assumption entails the validity of

$$c(x)^{p-1} - 1 | x^q - x.$$

On account of ($4_2$), the left-hand side is of degree $q-1$, whence (because $c(x)$ is monic)

$$c(x)^{p-1} - 1 = \frac{x^q - x}{x - \varkappa}$$

follows for suitable $\varkappa \in \mathsf{F}$. Thus

$$c(x - \varkappa)^{p-1} - 1 = \frac{x^q - x}{x} = x^{q-1} - 1;$$

therefore

$$c(x - \varkappa) = x^{\frac{q-1}{p-1}}$$

holds, whence (3) follows. This proves the assertion.

Fortunately it will turn out, *inter alia*, as we have already suggested, that the regular $c(x)$ are friendly towards both values $\tau = 0$, 1, i.e. they yield solutions of (the marginal case) of Problem II. For this reason we shall call these solutions *regular*. They can be given explicitly. Moreover, and this is a much stronger assertion, it is even true that, apart from $x^q - x$, Problem II has no further solutions whatsoever, with the exception of at most $q = 3^n$, $5^n$, $7^n$ ($n \geqq 4$). More precisely, we prove

THEOREM 17. *Problem II, consisting—we repeat—in the determination of the fully reducible $f(x)$ such that*

$$f(x) \in \mathsf{F}[x] \setminus \mathsf{F}[x^p], \quad f^{\circ} = q, \quad f^{\circ\circ} \leqq \frac{q+1}{2}, \tag{6}$$

*has, apart from the trivial solution $x^q - x$ only solutions belonging to the marginal case $f^{\circ\circ} = \frac{q+1}{2}$ $(p \neq 2)$. The regular solutions $f(x)$ are obtained by taking an element*

$$\sigma = \pm 1 \qquad (\in \mathsf{F}_p), \tag{7}$$

*natural numbers $P_0, \ldots, P_k$ such that*

$$P_0 \underset{1}{|} P_1 \underset{1}{|} \cdots \underset{1}{|} P_k, \qquad p = P_0 < P_1 < \cdots < P_k = q, \tag{8}$$

*and elements $a_0, \ldots, a_{k-1}$ with the quadratic character values*

$$\chi(a_i) = 0, \sigma \qquad (i = 0, \ldots, k-1) \tag{9}$$

*and, finally, elements $\varrho_i \in \mathsf{F}_{P_i}$ $(i=0, \ldots, k)$, such that $\varrho_0, \ldots, \varrho_{k-1}$ are given by the recursive formula*

$$\varrho_0 = a_0, \quad \varrho_i = \Big(\Big(\ldots(a_i \ldots \varrho_0)^{\frac{P_0-1}{P_1-1}} - \varrho_{i-1}\Big)\ldots\Big)^{\frac{P_{i-2}-1}{P_{i-1}-1}} - \varrho_{i-1}\Big)^{\frac{P_{i-1}-1}{P_i-1}}, \tag{10}$$

*while $\varrho_k$ is allowed to be an arbitrary element of $\mathsf{F}_{P_k}$ $(=\mathsf{F})$, then forming the polynomial*

$$c(x) = \Big(\ldots\Big((x+\varrho_k)^{\frac{P_k-1}{P_{k-1}-1}} + \varrho_{k-1}\Big)\ldots\Big)^{\frac{P_1-1}{P_0-1}} + \varrho_0 \tag{11}$$

*with these elements $\varrho_0, \ldots, \varrho_k$ and putting*

$$f(x) = \frac{x^q - x}{c(x)^{\frac{p-1}{2}} - \sigma} \Big(c(x)^{\frac{p-1}{2}} - \sigma\tau\Big) \qquad (\tau = 0, 1). \tag{12}$$

*Apart from the cases*

$$q = 3^n, 5^n, 7^n \qquad (n \geqq 4) \tag{13}$$

*Problem II has no other solutions.*

SUPPLEMENT. *The above holds even under the restriction*

$$\varrho_1, \ldots, \varrho_{k-1} \neq 0, \tag{14}$$

*which makes the described construction unique.*

Before giving the proof we shall present three simple examples, which will also help to explain the theorem.

EXAMPLE 1. Let $\mathsf{F}$ be the prime field $\mathsf{F}_p$ (i.e. $q=p$, $n=1$). In this case (and only in this case) the sequence $P_0, \ldots, P_k$ consists of the single term $P_0(=p)$ (i.e. in this case $k=0$, while in all other cases $k \geqq 1$.) The elements $a_i$ are now missing, and the elements $\varrho_i$ reduce to the single element $\varrho_0$ which can now be chosen arbitrarily from $\mathsf{F}_p$. (11) is to be understood simply as $c(x) = x - \varrho_0$. Replacing $\varrho_0$ by $a$ we obtain

$$f(x) = \frac{x^q - x}{(x+a)^{\frac{p-1}{2}} + \sigma}\left((x+a)^{\frac{p-1}{2}} - \sigma\tau\right)$$

from (12). For the numerator let us write

$$(x+a)^p - (x+a).$$

We get

$$f(x) = (x+a)\left((x+a)^{\frac{p-1}{2}} - \sigma\right)\left((x+a)^{\frac{p-1}{2}} - \sigma\tau\right) \qquad (\sigma = \pm 1;\ \tau = 0, 1). \quad (15)$$

Accordingly, this $f(x)$ and $x^p - x$ are the only solutions of Problem II for $\mathsf{F} = \mathsf{F}_p$ $(p \neq 2)$. This corrects a (common) error and is due to Rédei [6, 12] (see also Rédei [13]).

In the next two examples we shall merely calculate (11) without substituting into (12).

Example 2. Let the prime number $n$ be the degree of $\mathsf{F}$. (The exceptions (13) do not apply to the quadratic and cubic cases, $n=2$ and $n=3$.) The sequence $P_0, \ldots, P_k$ must consist of the terms $P_0 = p$, $P_1 = q$ (so we have $k=1$). The elements $a_i$ now reduce to the single element $a_0 \in \mathsf{F}_p$ which we shall write as $a$. This is characterized by

$$a \in \mathsf{F}_p, \quad \chi(a) = 0, \varrho.$$

The elements $\varrho_i$ reduce to $\varrho_0 = a$ and an arbitrary $\varrho_1 = \varrho$ from $\mathsf{F}$. We have

$$c(x) = (x+\varrho)^{\frac{q-1}{p-1}} + a.$$

Example 3. Let $n$, the degree of $\mathsf{F}$, be a product of two (equal or different) prime numbers $r, s$, i.e. $n = rs$. We have the following three cases for $P_0, \ldots, P_k$:

1) $P_0 = p, \quad P_1 = q \quad (k=1),$
2) $P_0 = 1, \quad P_1 = p^r, \quad P_2 = q \quad (k=2),$
3) $P_0 = 1, \quad P_1 = p^s, \quad P_2 = q \quad (k=2)$

(the last two of these are the same if $r=s$). We content ourselves with a more detailed explanation of case 2). We take elements $a_0, a_1 \in \mathsf{F}_p$ such that

$$\chi(a_0) = 0, \sigma; \quad \chi(a_1) = 0, \sigma$$

and elements $\varrho_i \in \mathsf{F}_{P_i}$ $(i = 0, 1, 2)$ such that

$$\varrho_0 = a_0, \quad \varrho_1 = \sqrt[\frac{P_1-1}{p-1}]{a_1 - a_0}.$$

From (11)

$$c(x) = \left((x+\varrho_2)^{\frac{q-1}{P_1-1}}+\varrho_1\right)^{\frac{P_1-1}{p-1}}+\varrho_0.$$

Substitution of this $c(x)$ into (12) leads, by (14), to solutions $f(x)$ that do not already belong to case 1) if and only if $\varrho_1 \neq 0$, i.e. $a_1 \neq a$.

We prove Theorem 17 as follows. We disregard the trivial solution $x^q - x$ of Problem II. As before, we denote by $c(x)$ one of the polynomials defined by (4) and (5). Taking into consideration the consequences of the Main Lemma given in (1) and (2), it suffices to prove these three assertions:

1) The regular $c(x)$ agree with the polynomials (11).

2) Every regular $c(x)$ is friendly towards both values $\tau = 0, 1$.

3) Apart from the cases (13), every $c(x)$ friendly towards $\tau=0$ or $\tau=1$ is regular.

Let us apply Theorem 11 to the special case (2) and use the fact that in this case the cofactor $\bar{\varphi}(x)$ of $\varphi(x)$ is determined by

$$\bar{\varphi}(x) = \frac{x^p - x}{\varphi(x)} = x\left(x^{\frac{p-1}{2}} - \sigma\right).$$

The zeros $\xi$ (lying in $\mathsf{F}_p$) of $\bar{\varphi}(x)$ are therefore characterized by

$$\chi(\xi)=0 \quad \text{or} \quad \chi(\xi)=\sigma.$$

This establishes the validity of the assertion 1). For this reason the $c(x)$ can henceforth also be defined from (7)—(11).

The assertion 2) states that for a $c(x)$ (defined by (7)—(11)) both polynomials

$$c(x)^{\frac{p-1}{2}} - \sigma\tau \qquad (\tau = 0, 1)$$

are fully reducible. (In case (3) this has already been proved but the following proof also covers this case.) We shall prove that moreover

$$c(x)^p - c(x)$$

is fully reducible. This means that every

$$c(x) - \omega \qquad (\omega \in \mathsf{F}_p)$$

is fully reducible. It is therefore sufficient to prove that for every element $\xi$ of the algebraic closure $\overline{\mathsf{F}}$ the assumption

$$c(\xi) \in \mathsf{F}_p \tag{16}$$

implies the validity of $\xi \in \mathsf{F}$.

To prove this, we write (11) in the form

$$c(x) = e_0(e_1(\cdots(e_{k-1}(x+\varrho_k))\dots))$$

with the notation

$$e_i(x) = x^{\frac{P_{i+1}-1}{P_i-1}} + \varrho_i \qquad (i = 0, \dots, k-1)$$

as used already. Then assumption (16) runs as follows:

$$e_0(e_1(\dots(e_{k-1}(\xi))\dots)) \in \mathsf{F}_p$$

after the allowed substitution $\xi + \varrho_k \rightarrow \xi$.

We prove more generally that for every $i=1, \dots, k$ and all $\xi \in \bar{\mathsf{F}}$ the validity of $\xi \in \mathsf{F}_{P_i}$ follows from the assumption

$$e_0(e_1(\dots(e_{i-1}(\xi))\dots)) \in \mathsf{F}_p. \tag{17}$$

For $i=1$ (17) reads $e_0(\xi) \in \mathsf{F}_p$, i.e. (because $P_0=p$)

$$\xi^{\frac{P_1-1}{p-1}} + \varrho_0 \in \mathsf{F}_p.$$

From (10), the second term of the left-hand side lies in $\mathsf{F}_p$ and so it can be cancelled. The assertion then follows trivially for this case. In the remaining case $i \geqq 2$ we assume the validity of the assertion for $i-1$. By this induction hypothesis, it follows from (17) that $e_{i-1}(\xi) \in \mathsf{F}_{P_{i-1}}$, i.e. that

$$\xi^{\frac{P_i-1}{P_{i-1}-1}} + \varrho_{i-1} \in \mathsf{F}_{p_{i-1}}.$$

By $(10_2)$, the second term of the left-hand side lies in the right-hand side, so that $\xi \in \mathsf{F}_{P_i}$. Thus we have proved assertion 2).

In the proof of assertion 3) we shall distinguish the three cases ($n \leqq 2$, $n \leqq 3$, $n \leqq 4$), i.e.

$$q \leqq p^2, \quad q = p^3, \quad q \geqq p^4.$$

According to Theorem 13, in the first case $c(x)$ is linear or non-primitive (with a linear successor), and therefore it is regular. In the second case the regularity of $c(x)$ follows from the corollary of Theorem 15, apart from the characteristics $p=3, 5$, which we shall investigate separately. In the third case the same follows from the corollary of Theorem 16, apart from (13). (We must emphasize that the friendliness of $c(x)$ has not yet been exploited.) If, in addition, we show that the above

$$q = 3^3, 5^3$$

are not really exceptions (for the friendly $c(x)$), then assertion 3) and thus also Theorem 17 will have been proved.

The proof of the few remaining assertions will be relatively difficult but quite interesting. For the sake of clarity we shall consider the case

$$q = p^3 \qquad (p \neq 2) \tag{18}$$

in complete generality and prove that then every $c(x)$ which is friendly towards $\tau^2=0$ or $\tau^2=0$, is regular, although we need only the cases $p=3$, 5. (Viz., for $p \geqq 7$ the question will be only that of a second proof of known facts.)

We need the following

EXAMPLE 4. For the case $q=p^3$ $(p \neq 2)$ we shall determine all the fully reducible polynomials of the form

$$g_\varkappa(x) = x^{p^2+p+1} + \varkappa(x^{p^2} + x^p + x + \lambda) \qquad (\varkappa \in \mathsf{F}_p \diagdown 0;\ \lambda \in \mathsf{F}_p). \tag{19}$$

(In reality $g_\varkappa(x)$ also depends on $\lambda$.) We form the product

$$G(x) = \prod_{\varkappa \in \mathsf{F}_p \diagdown 0} g_\varkappa(x).$$

The right-hand side is equal to

$$x^{p^3-1} - (x^{p^2} + x^p + x + \lambda)^{p-1},$$

so that (because $p^3 = q$)

$$(x^{p^2} + x^p + x + \lambda)G(x) = (x^{p^2} + x^p + x + \lambda)x^{q-1} - (x^{p^2} + x^p + x + \lambda)^p,$$

i.e. after a few minor adjustments, using $\lambda^p = \lambda$,

$$(x^{p^2} + x^p + x + \lambda)G(x) = (x^{q-1} - 1)(x^p + x + \lambda)^p.$$

Hence

$$g_\varkappa(x) | (x^{q-1} - 1)(x^p + x + \lambda)^p.$$

As the first factor on the right-hand side is fully reducible, $g_\varkappa(x)$ is fully reducible if and only if the greatest common divisor

$$\Delta(x) = (g_\varkappa(x),\ x^p + x + \lambda)$$

is fully reducible. Now,

$$\Delta(x) = (x^{p^2+p+1} + \varkappa(x^{p^2} + x^p + x + \lambda,\ x^p + x + \lambda) = (x^{p^2+p+1} + \varkappa x^{p^2},\ x^p + x + \lambda).$$

Thus $\Delta(x)$ is fully reducible if and only if the same holds for

$$\Delta_0(x) = (x^{p+1} + \varkappa,\ x^p + x + \lambda).$$

Since, however,

$$\Delta_0(x) = (x^2 + \lambda x - \varkappa,\ x^p + x + \lambda),$$

we infer that

$$\Delta_0\left(x - \frac{\lambda}{2}\right) = \left(x^2 - \frac{1}{4}(\lambda^2 + 4\varkappa),\ x^p + x\right).$$

Accordingly, $g_\varkappa(x)$ is fully reducible if and only if

$$\Delta_{00}(x) = \left(x^2 - \frac{1}{4}(\lambda^2+4\varkappa),\ x^{p-1}+1\right)$$

is fully reducible. By virtue of $2|p-1$ the other form

$$\Delta_{00}(x) = \left(x^2 - \frac{1}{4}(\lambda^2+4\varkappa),\ \left(\frac{1}{4}(\lambda^2+4\varkappa)\right)^{\frac{p-1}{2}}+1\right) =$$

$$= \left(x^2 - \frac{1}{4}(\lambda^2+4\varkappa),\ \chi(\lambda^2+4\varkappa)+1\right)$$

holds. Thus we have the following answer to our question: $g_\varkappa(x)$ is fully reducible if and only if

$$\chi(\lambda^2+4\varkappa) \geqq 0.$$

Now, to prove the above assertion, let us consider for the case (18) a $c(x)$, friendly towards a $\tau(=0, 1)$. We have to show that $c(x)$ is regular. As $n=3$ it is sufficient to show that $c(x)$ is non-primitive. With this aim, let as assume that $c(x)$ is primitive. We must deduce a contradiction from this assumption.

Of course, $c(x)$ can be subjected to a translation $x \to x+\xi$ ($\xi \in \mathsf{F}$). We may therefore restrict ourselves to the case $c^\downarrow \geqq 2$. As $c(x)$ is, in the first place, a stem polynomial for $p$, Theorem 3 implies that

$$c(x) = x^{p^2+p+1} + \alpha x + \alpha^p x^p + \alpha^{p^2} x^{p^2} + \beta \qquad (\alpha \in \mathsf{F},\ \beta \in \mathsf{F}_p). \tag{21}$$

The assumption that $c(x)$ is a solution of Problem III, belonging to the pair (2), then means, by Proposition 20, that

$$c_1(x) | c_0(x)^{\frac{p+1}{2}} - \sigma c_0(x), \tag{22}$$

where (by (21))

$$c_0(x) = \alpha^p x^p + \alpha x + \beta, \quad c_1(x) = x^{p+1} + \alpha^{p^2} \tag{23}$$

are the one-step constituents of $c(x)$. If we introduce the two-step constituents

$$c_{00}(x) = \alpha x + \beta, \quad c_{01} = \alpha^p, \quad c_{10}(x) = \alpha^{p^2}, \quad c_{11}(x) = x \tag{24}$$

and their determinant

$$D(x) = \begin{vmatrix} c_{00}(x) & c_{01}(x) \\ c_{10}(x) & c_{11}(x) \end{vmatrix} = \alpha x^2 + \beta x - \alpha^{p^2+p} \tag{25}$$

then the (necessary) condition

$$c_1(x) | D(x)^{\frac{p+1}{2}} - \sigma c_{11}(x)^{\frac{p-1}{2}} D(x) \tag{26}$$

also holds, by reason of Proposition 23. Finally, the assumption that $c(x)$ is friendly relative to a $\tau(=0, 1)$ means that

$$c(x)^{\frac{p-1}{2}} - \sigma\tau \tag{27}$$

is fully reducible.

As $c(x)$ is primitive,

$$\alpha \neq 0 \tag{28}$$

follows from (21).

By (25) and (28) we have $D^\circ = 2$, so that both sides of (26) are of degree $p+1$, and thus (26) (using $(23_2)$ and (25)) becomes

$$\alpha^{\frac{p+1}{2}} c_1(x) = D(x)^{\frac{p+1}{2}} - \sigma c_{11}(x)^{\frac{p-1}{2}} D(x).$$

After substituting from $(23_2)$, $(24_4)$ and (25) we have

$$\alpha^{\frac{p+1}{2}} (x^{p+1} + \alpha^{p^2}) = (\alpha x^2 + \beta x - \alpha^{p^2+p})^{\frac{p+1}{2}} - \sigma x^{\frac{p-1}{2}} (\alpha x^2 + \beta x - \alpha^{p^2+p}). \tag{29}$$

We first consider the case $p=3$. By (29),

$$\alpha^2 (x^4 + \alpha^9) = (\alpha x^2 + \beta x - \alpha^{12})^2 - \sigma x(\alpha x^2 + \beta x - \alpha^{12}).$$

Comparison of the coefficients of $x^0$ and $x^1$ gives the necessary conditions

$$\alpha^{11} = \alpha^{24}, \quad 0 = \alpha^{12}\beta + \sigma\alpha^{12},$$

i.e. (by (28))

$$\alpha^{13} = 1, \quad \beta = -\sigma. \tag{30}$$

Hence, and by (21), we have

$$c(x) = x^{13} + \alpha x + \alpha^3 x^3 + \alpha^9 x^9 - \sigma. \tag{31}$$

Therefore (cf. (27))

$$x^{13} + \alpha x + \alpha^3 x^3 + \alpha^9 x^9 - \sigma - \sigma\tau$$

is a fully reducible polynomial. The substitution $x \to \dfrac{x}{\alpha}$ gives us (from $(30_1)$) that

$$x^{13} + x^9 + x^3 + x - \sigma(1+\tau)$$

is also fully reducible. Now Example 4 can be used with $p=3$, $\varkappa=1$, $\lambda = -\sigma(1+\tau)$. It follows from (20) that

$$\chi((1+\tau)^2 + 4) \geqq 0,$$

i.e. $\chi((1+\tau)^2+1) \geqq 0$. This, however, contains a contradiction for both values $\tau=0, 1$.

Now let us consider the case $p \geqq 5$. The coefficient of $x^p$ on the right-hand side of (29) must vanish, and consequently $\beta=0$, by (28). Accordingly, (29) becomes

$$\alpha^{\frac{p+1}{2}}(x^{p+1}+\alpha^{p^2}) = (\alpha x^2-\alpha^{p^2+p})^{\frac{p+1}{2}} - \sigma x^{\frac{p-1}{2}}(\alpha x^2-\alpha^{p^2+p}).$$

After expanding, the series on the right-hand side consists of $\frac{p+3}{2}$ (non-vanishing) terms. The number of terms remaining in this equation is $\leqq 4$, so that $\frac{p+3}{2} \leqq 4$ (the $=$ sign must in fact hold here). It follows that $p \leqq 5$, and thus $p=5$. We have

$$\alpha^3(x^6+\alpha^{25}) = (\alpha x^2-\alpha^{30})^3 - \sigma x^2(\alpha x^2-\alpha^{30}).$$

From the substitution $x^2 \to x$ we get

$$\alpha^3(x^3+\alpha^{25}) = (\alpha x-\alpha^{30})^3 - \sigma x(\alpha x-\alpha^{30}).$$

Comparison of the coefficients of $x^1$ gives

$$0 = 3\alpha^{61}+\sigma\alpha^{30},$$

i.e. (as $p=5$)

$$\alpha^{31} = -2\sigma. \tag{32}$$

Now (cf. (21) and (27))

$$\big(c(x)^2-\sigma\tau =\big)(x^{31}+\alpha^{25}x^{25}+\alpha^5x^5+\alpha x)^2-\sigma\tau$$

must be fully reducible. The same also holds after the substitution $x \to \frac{x}{\alpha}$, consequently (using (32) and $(2\sigma)^2=-1$) also

$$(x^{31}-2\sigma(x^{25}+x^5+x))^2+\sigma\tau \tag{33}$$

is fully reducible. As can easily be seen, this polynomial can be decomposed into the product

$$\prod_{e=1,\,-1}(x^{31}-2\sigma(x^{25}+x^5+x+e(\sigma+3)\tau));$$

thus both factors are fully reducible. Application of Example 4 with

$$p = 5, \quad \varkappa = -2\sigma, \quad \lambda = e(\sigma+3)\tau$$

gives (by virtue of $e=\pm 1$, $\sigma=\pm 1$, $\tau=0, 1$)

$$\big(\chi(\lambda^2+4\varkappa) =\big)\chi(\sigma\tau+2\sigma) \geqq 0,$$

i.e.

$$\chi(\tau+2) \geqq 0.$$

Since this is false for both values $\tau=0, 1$, this contradiction completes the proof of Theorem 17.

In order to prove the supplement, let us write (12) as follows:

$$f(x) = x^q - x - \sigma(1+\tau)\frac{x^q - x}{c(x)^{\frac{p-1}{2}} + \sigma}.$$

Hence we see that $\sigma$, $\tau$, $c(x)^{\frac{p-1}{2}}$ and even $c(x)$ are uniquely determined by $f(x)$ (where we have used the fact that $c(x)$ is monic). Consequently the validity of the supplement follows from the final assertion of Theorem 11.

NOTE 1. For $p=3$ in case (30), (31) satisfies condition (22) (cf. (23)) and

$$c(x) = x^{31} + \alpha x + \alpha^5 x^5 + \alpha^{25} x^{25}$$

similarly satisfies condition (22) (cf. (23)) for $p=5$ in case (32), as is easily seen from a simple calculation. This means that in the cases $q=3^3, 5^3$ the polynomials

$$(x+\xi)^{13} + \alpha(x+\xi) + \alpha^3(x+\xi)^3 + \alpha^9(x+\xi)^9 - \sigma \qquad (\alpha, \xi \in \mathsf{F};\ \alpha^{13} = 1)$$

and

$$(x+\xi)^{31} + \alpha(x+\xi) + \alpha^5(x+\xi)^5 + \alpha^{25}(x+\xi)^{25} \qquad (\alpha, \xi \in \mathsf{F};\ \alpha^{31} = -2\sigma)$$

are primitive (and thus non-regular) solutions of Problem III, belonging to the pair $(P=)3$, $(\varphi(x)=)x+\sigma$ or to the pair $(P=)5$, $(\varphi(x)=)x^2+\sigma (\sigma=\pm 1)$, respectively. (Nevertheless, according to the above, these solutions of Problem III do not lead to any solutions of Problem II.)

NOTE 2. As the proof of Theorem 17 for $q=3^3, 5^3$ was relatively difficult, there is not much hope that, without a new idea, the exceptional cases $q=3^n$, $5^n$, $7^n$ $(n \geqq 4)$ might be amenable to similar investigation. Our attempts to do this have already failed in the "easiest" cases $q=3^4, 5^4, 7^4$.

## § 33. Appendix on the degenerate solutions of Problem II

In Problem II we dealt with fully reducible polynomials $f(x)$ such that

$$f(x) \underset{1}{\in} \mathsf{F}[x], \quad f^{\circ} = q, \quad f^{\circ\circ} \leqq \frac{q+1}{2},$$

where the degenerate case $f'(x)=0$, i.e.

$$f(x) \in \mathsf{F}[x^p]$$

was disregarded. We now proceed to deal with this case. As already mentioned, the device applied in the non-degenerate case now fails. It will, however, be applicable (in a slightly modified form) after it has been subjected to a simple transformation. In this way we shall obtain a rather more modest result, which nevertheless extends Theorem 17 and is important for its applications.

If $p^e$ $(1 \leqq e \leqq n)$ is the exponent of the above $f(x)$, then we can put

$$f(x) = g(x)^{p^e}, \quad g(x) \underset{1}{\in} \mathsf{F}[x] \setminus \mathsf{F}[x^p], \quad g^\circ = \frac{q}{p^e} \quad \left( g^{\circ\circ} \leqq \frac{q+1}{2p^e} \right),$$

where not only $f(x)$ but also $g(x)$ must be fully reducible. This is the required transformation. The trivial case $e=n$ can be disregarded; therefore we may assume that $n \geqq 2$. The best attainable result is then an estimation from below for $g^{\circ\circ}$, the second degree of the solutions. More precisely, the following theorem holds (with appropriate changes in the notation):

THEOREM 18. *If* ($n \geqq 2$ *and*) $e \left( 1 \leqq e \leqq \frac{n}{2} \right)$ *is a natural number and* $f(x)$ *is a fully reducible polynomial such that*

$$f(x) \underset{1}{\in} \mathsf{F}[x] \setminus \mathsf{F}[x^p], \quad f^\circ = \frac{q}{p^e}, \tag{1}$$

*then*

$$f^{\circ\circ} \geqq \frac{q+p^e}{p^e(p^e+1)} \left( = \frac{p^{n-e}+1}{p^e+1} \right) \tag{2}$$

*must hold.*

(This theorem also holds, trivially, for $\frac{n}{2} < e < n$, as then the right-hand side of (2) is less than 1. It holds, as a consequence of Theorem 17, also for $n \geqq 1$, $e=0$ (and thus, finally, for all $0 \leqq e < n$) with the exception of $f(x) = x^q - x$.)

To prove Theorem 18 we use the zero polynomial

$$F(x) = \prod_{f(\varrho)=0} (x-\varrho)$$

of $f(x)$ (as was also done in connection with Problem II). Since $f(x)$ is fully reducible, it follows that

$$F(x) | x^q - x.$$

Now, again,

$$F(x) | f(x), \quad \frac{f(x)}{F(x)} \bigg| f'(x)$$

must hold, trivially.

By virtue of (1) we can put

$$f(x) = x^{\frac{q}{p^e}} + g(x), \quad g(x) \in \mathsf{F}[x] \setminus \mathsf{F}[x^p], \tag{3}$$

whence

$$f'(x) = g'(x) \neq 0.$$

On the one hand we obtain

$$F(x) | f(x)^{p^e},$$

i.e.

$$F(x) | x^q + g(x)^{p^e},$$

so that

$$F(x) | x + g(x)^{p^e}.$$

On the other hand we have

$$\left.\frac{f(x)}{F(x)}\right| g'(x).$$

By multiplying the two divisibilities we get

$$f(x) | (g(x)^{p^e} + x) g'(x). \tag{4}$$

Thus, as $g^{\circ} \geqq 1$, comparison of the coefficients leads to

$$(p^e + 1) g^{\circ} - 1 \geqq (f^{\circ} =) \frac{q}{p^e},$$

from which Theorem 18 follows.

NOTE. There is in connection with this theorem a marginal problem consisting of the investigation of the $f(x)$ for which (2) is valid with "=". It is necessary, for this to be so, that the right-hand side of (2) should be an integer, i.e. that

$$2e | n$$

should hold. A further necessary condition can be deduced as follows. If the marginal case occurs, then the two sides of (4) must be associated with one another. This implies the validity of

$$\gamma f(x) = (g(x)^{p^e} + x) g'(x)$$

with a suitable $\gamma \in \mathsf{F} \setminus 0$. In view of (3), we obtain by differentiation that

$$\gamma g'(x) = g'(x) + (g(x)^{p^e} + x) g''(x),$$

whence clearly $g''(x) = 0$, $\gamma = 1$ and also $g'(x) \in \mathsf{F}[x^p]$. We have not had further success in our attempts to determine the possible $f(x)$.

EXAMPLE. In the extreme case

$$n = 2e$$

there exist functions $f(x)$ from Theorem 18 for which (2) holds with "=". All such $f(x)$ can be determined as follows. As $q=p^{2e}$, $f^{\circ\circ}=1$ must now hold. This means that

$$f(x) = x^P + \alpha x + \beta \qquad (\alpha \in \mathsf{F}\setminus 0,\ \beta \in \mathsf{F}), \tag{5}$$

where

$$P = p^e (q = P^2)$$

has been put for brevity. It remains only to determine which are the fully reducible ones among these $f(x)$. Because $f'(x) = \alpha \neq 0$, we have

$$f(x) | x^q - x$$

so that by $f(x)^P = x^q + \alpha^P x^P + \beta^P$,

$$f(x) | x + \alpha^P x^P + \beta^P$$

follows. Thus the right-hand side must be equal to $\alpha^P f(x)$. From (5), this implies that

$$\alpha^{P+1} = 1, \quad \alpha^P \beta = \beta^P.$$

Accordingly, only the two cases

$$\alpha^{P+1} = 1, \quad \beta = 0 \quad \text{or} \quad \alpha = \beta^{1-P} \neq 0 \tag{6}$$

need be considered. In the first case we have $\alpha = \gamma^{P-1}$ with $\gamma \in \mathsf{F}\setminus 0$ and therefore (see (5)) the corresponding

$$f(x) = x^P + \gamma^{P-1} x$$

are clearly fully reducible. In the second case we have

$$f(x) = x^P + \beta^{1-P} x + \beta.$$

We shall show that this $f(x)$, too, is fully reducible. If suffices to prove that

$$g(x) = \beta^P f\left(\frac{x}{\beta}\right) = x^P + x + \beta^{P+1}$$

is fully reducible, i.e. it has $P$ zeros in $\mathsf{F}$. For this purpose we write the elements of $\mathsf{F}$ in the form

$$\omega = \varrho\omega_1 + \sigma\omega_2 \qquad (\varrho, \sigma \in \mathsf{F}_P)$$

where $\omega_1, \omega_2$ is a normal basis of $\mathsf{F}$ relative to $\mathsf{F}_P$ ($\omega_1^P = \omega_2$). Now,

$$g(\omega) = (\varrho + \sigma)(\omega_1 + \omega_2) + \beta^{P+1},$$

so that

$$\omega_1 + \omega_2 \in \mathsf{F}_P\setminus 0, \quad \beta^{P+1} \in \mathsf{F}_P$$

means that $g(\omega)=0$ holds if and only if

$$\varrho+\sigma = -\frac{\beta^{P+1}}{\omega_1+\omega_2}$$

whence it follows that $P$ different zeros of $g(x)$, consequently also of $f(x)$, lie in $\mathsf{F}_P$. Thus we have shown that for $n=2e$ all the $f(x)$ belonging to the marginal case of Theorem 19 are given by (5) and (6). Of course, this example shows also that the lower bound given in (2) is in general exact.

CHAPTER VI

# APPLICATIONS

The results obtained for Problems I and II, i.e. Theorems 5, 17 and 18, lead, as was pointed out in the Preface, to surprisingly diverse and far-reaching applications. Each of sections 34—39 deals with a different domain of application. These applications have a particular fascination as, in all probability, no other approach can be found for them. For, the devices used to prove theorems 5, 17 and 18 have in each case caused a decisive change and marked progress, and these changes can, without doubt, only be formulated without constraint in the "polynomial language". We feel that, if we neglect these theorems we are necessarily also deprived of their applications (see, however, § 38, Note 3). Unfortunately, for the marginal case of Theorem 17, only in the case of a prime field $q=p$ did we succeed in making practical use of the theorem. It is a meagre consolation that no practical applications whatsoever would have been possible on the strength of the marginal theorem of Minkowski–Hajós, so that, as far as applications of our marginal theorem are concerned, we are, for the moment, slightly better off.

## § 34. Certain families of linear mappings in finite fields

First of all let S denote an arbitrary non-empty set and let $f$ be a function $f\colon \mathsf{S}\times\mathsf{S}\rightarrow\mathsf{S}$. Then for every fixed $\eta$

$$\xi\rightarrow f(\xi,\eta)$$

is a mapping of S into itself and the same thing holds for

$$\eta\rightarrow f(\xi,\eta)$$

for every fixed $\xi$. We say then that we have two *dual families* of mappings of S into itself. For the first family $\eta$ and for the second $\xi$ play the role of a *parameter of the family*. The reader mey see that we are dealing here with concepts that will often reappear in various connections. In order to make the situation clearer, let $\xi_i$ and $\eta_i$ run through all elements of S for (finitely or infinitely many) indices $i=1, 2, \ldots$ and then form the matrix

$$\begin{matrix} f(\xi_1,\eta_1) & f(\xi_2,\eta_1)\ldots \\ f(\xi_1,\eta_2) & f(\xi_2,\eta_2)\ldots \\ \vdots & \vdots \end{matrix}$$

The rows and columns give the first and second families of mappings respectively. The order of the rows and that of the columns are irrelevant.

Of course, two such (dual) families are each determined by the other. It is for just this reason interesting to try to discover how else they are connected. In other terms, what properties of one of the families can be inferred from those of the other?

We intend to investigate the case in which the given set coincides with our finite field F and the first family is taken as linear. Then the question (*inter alia*) of the number of permutations appearing in the second family arises quite naturally. The chief result in this connection is:

THEOREM 19. *For q linear polynomials*

$$l_\eta(x) \in \mathsf{F}[x] \qquad (\eta \in \mathsf{F},\ l_\eta^\circ \leqq 1) \tag{1}$$

*the number* $(0 \leqq N \leqq q)$ *of permutations appearing among the q mappings*

$$\eta \to l_\eta(\xi) \qquad (\xi \in \mathsf{F}) \tag{2}$$

*lies in one of the intervals*

$$\left(0, \frac{q-1}{2}\right);\ \left(q - \frac{q-1}{p^e - 1},\ q - 1 - \frac{q-1}{p^e+1}\right)\ \left(e = 1, \ldots, \left[\frac{n}{2}\right]\right);\ (q-1, q) \tag{3}$$

(*clearly pairwise disjoint*).

(We must emphasize that we treat constants as linear polynomials. Accordingly, the mappings of F into itself with a single image element are treated as linear mappings. Of course, the polynomials (1) must be distinct when $N > 0$.)

The reader will notice that in fact Theorem 19 deals with dual families of mappings of F into itself, where the first family consists of the mappings (not given particular emphasis in the text)

$$\xi \to l_\eta(\xi) \qquad (\eta \in \mathsf{F}), \tag{4}$$

while the second is given by the mappings (2).

$N$ can be regarded as a function of system (1). By virtue of Theorem 19, this function is (in general) strongly "lacunary". Only relatively few of its distinct values can lie in the second half of the interval $(0, q)$ for large values of $p$. One should bear in mind that the intervals (3) are arranged from left to right on the number line, their lengths being respectively:

$$\frac{q-1}{2},\ \frac{2(q-1)}{p^2-1},\ \frac{2(q-1)}{p^4-1},\ \ldots,\ \frac{2(q-1)}{p^{2\left[\frac{n}{2}\right]}-1},\ 2.$$

CONJECTURE. $N$ assumes every integer value that lies in one of the intervals (3).

EXAMPLE 1. For $q=p$, $N$ lies in one of the two intervals

$$\left(0, \frac{p-1}{2}\right), \quad (p-1, p).$$

This result was obtained by Rédei [6, 12] but the proof was based on an error (see also Rédei [13]).

EXAMPLE 2. For $q=p^2$, $N$ lies in one of the three intervals

$$\left(0, \frac{p^2-1}{2}\right), \quad (p^2-p-1, p^2-p), \quad (p^2-1, p^2).$$

EXAMPLE 3. For $q=p^3$, $N$ lies in one of the three intervals

$$\left(0, \frac{p^3-1}{2}\right), \quad (p^3-p^2-p-1, p^3-p^2+p-2), \quad (p^3-1, p^2)$$

In addition to Theorem 19, we shall establish further results for the mappings (2) which will partly support the above conjecture and will form a necessary prelude to the proof of Theorem 19. We shall not perform the proof in the shortest way, but shall first deal with one or two points which could equally well have been postponed. We note that the mappings (2) are of interest not only in themselves but are also the source for almost all the investigations that we shall carry out in this chapter.

We shall stick to the notation introduced in Theorem 19. We shall also need the coefficient representation of the polynomials (1) which we shall write in the form

$$l_\eta(x) = \alpha_\eta x + \beta_\eta \qquad (\eta \in \mathsf{F};\ \alpha_\eta,\ \beta_\eta \in \mathsf{F}). \tag{5}$$

The mappings (2) now assume the more complete form

$$\eta \to l_\eta(\xi) = \alpha_\eta \xi + \beta_\eta \qquad (\xi \in \mathsf{F}). \tag{6}$$

In a few places we shall need the usual extension

$$\mathsf{F}^* = \mathsf{F} \cup \infty$$

of $\mathsf{F}$ with the *infinite element* $\infty$. We shall attach to the latter (from (6)) the mapping

$$\eta \to l_\eta(\infty) = \alpha_\eta \tag{7}$$

(of $\mathsf{F}$ into itself), determined likewise by system (1). (As indicated, we define the substitution value of $l_\eta(x)$ at the "place" $\infty$ by $l_\eta(\infty)=\alpha_\eta$.) Taking (7) with the mappings (2), we obtain the $q+1$ mappings

$$\eta \to l_\eta(\xi) \qquad (\xi \in \mathsf{F}^*) \tag{8}$$

which we shall call the *extended mappings* belonging to (1). Their totality will be said to be the *extended dual family*. The number of permutations (of F) appearing in this family will be denoted by $N^*$. Thus $N^* = N+1$ if (7) is a permutation, otherwise $N^*=N$.

NOTE 1. We could alternatively have arrived at these concepts in the following way. Let us form the homogeneous polynomials

$$l_\eta(x, y) = \alpha_\eta x + \beta_\eta y$$

from the polynomials (1) and define the projective straight line $g$ over F consisting of the $q+1$ points

$$(\varkappa, \lambda) \qquad (\varkappa, \lambda \in \mathsf{F}; \quad \text{not} \quad \varkappa=\lambda=0),$$

where two points $(\varkappa_i, \lambda_i)$ $(i=1, 2)$ with

$$\begin{vmatrix} \varkappa_1 & \lambda_1 \\ \varkappa_2 & \lambda_2 \end{vmatrix} = 0$$

are to be considered to be equal, so that all points of $g$ can be written in the normal form

$$(\varkappa, 1) \ (\varkappa \in \mathsf{F}), \qquad (1, 0).$$

The (extended) mappings (9) are then the

$$\eta \to l_\eta(\varkappa_i, \lambda_i) \qquad (i = 1, \ldots, q+1)$$

where $(\varkappa_i, \lambda_i)$ runs through the points of $g$ (in their normal form). This fact will not be used in what follows.

For a mapping

$$\eta \to F(\eta)$$

of F into itself the product

$$\prod_{\eta \in \mathsf{F}} (x - F(\eta)) \qquad (\in \mathsf{F}[x])$$

is called its *accompanying polynomial*. This polynomial is fully reducible, monic and of degree $q$. Its zeros are precisely the images appearing in $\eta \to F(\eta)$, each one counted with the corresponding multiplicity. It is important that for the permutations, and only for these, the accompanying polynomial is equal to $x^q - x$. Another special case, also important for our purpose, arises when the accompanying polynomial has an exponent $p^e$ $(e \geqq 1)$, (i.e. it lies in

$\mathsf{F}[x^{p^e}]\setminus\mathsf{F}[x^{p^{e+1}}]$), in which case $\eta\to F(\eta)$ will be called a *$p^e$-mapping*. (Every image element thus appears in this mapping with a multiplicity divisible by $p^e$.)

For the accompanying polynomials of the mappings (2) we introduce the notation

$$f_\xi(x) = \prod_{\eta\in\mathsf{F}} (x - l_\eta(\xi)) \qquad (\xi\in\mathsf{F}). \tag{9}$$

These polynomials will play an important role in what follows.

Apart from $N$, we shall also have to deal with the number of $p^e$-mappings appearing in (2), written as $N_{p^e}$ ($1\leqq e\leqq n$). We remark that $N$ and $N_{p^e}$ can thus be defined as the number of accompanying polynomials $f_\xi(x)$ ($\xi\in\mathsf{F}$) that are equal to $x^q-x$ or are of exponent $p^e$, respectively.

As has recently become customary, we shall understand by the *kernel* of a mapping of $\mathsf{F}$ into itself the partition of $\mathsf{F}$ into classes so that exactly those elements with a common image belong to the same class. Then for $e\geqq 1$ the proposition "(2) is a $p^e$-mapping" is equivalent to the fact that the cardinalities of the classes of the kernel of (2) are multiples of $p^e$, $e$ being maximal for this property. The permutations of $\mathsf{F}$ are characterized by the fact that their kernel is trivial, which is to be understood as meaning that each of its classes contains one element.

Two sets are said to be *almost disjoint* if they contain at most one common element. Two classifications of a set are similarly called *almost disjoint* if the classes appearing in them are pairwise almost disjoint. (As a classical example, the bundles of parallel straight lines in affine space are a system of pairwise almost disjoint classifications of this space when space and straight line are interpreted as point sets.) We prove

PROPOSITION 36. *If $N>0$ the kernels of the mappings* (2) *are almost pairwise disjoint.*

If this were false, then there would be two different $\xi_1, \xi_2\in\mathsf{F}$ such that the kernels of the mappings

$$\eta\to l_\eta(\xi_i) \qquad (i=1,2)$$

would each contain a class in which two different common elements $\eta_1, \eta_2\in\mathsf{F}$ appear. This would mean that

$$l_{\eta_1}(\xi_i) = l_{\eta_2}(x) \qquad (i=1,2).$$

Hence, by putting

$$l(x) = l_{\eta_1}(x) - l_{\eta_2}(x),$$

$$l(\xi_i) = 0 \qquad (i=1,2)$$

follows; thus, certainly, $l(x)=0$, i.e.

$$l_{\eta_1}(x)=l_{\eta_2}(x),$$

in view of $\xi_1\neq\xi_2$. Accordingly, as $\eta_1\neq\eta_2$, no permutations at all occur in (2), but this is a contradiction of the fact that $N>0$. Thus we have proved Proposition 36.

PROPOSITION 37. *If all classes contain at least $k\geqq 2$ elements in pairwise almost disjoint classifications $C_1, \ldots, C_r$ of a non-void set $M$, then*

$$r \leqq \frac{O(M)-1}{k-1}. \tag{10}$$

To prove this, we take an element $a\in M$ and write $C_i(a)$ for the class of the $C_i$ $(i=1, \ldots, r)$, that are represented by $a$. By the assumption

$$C_i(a)\cap C_j(a)=a \qquad (i\neq j;\ i,j=1, \ldots, r) \tag{11}$$

and

$$O(C_i(a))\geqq k \qquad (i=1, \ldots, r). \tag{12}$$

From (11), the difference sets

$$C_i(a)\setminus a \qquad (i=1, \ldots, r)$$

are pairwise disjoint and therefore

$$\sum_{i=1}^{r}\left(O(C_i(a))-1\right) \leqq O(M)-1.$$

Hence and from (12)

$$r(k-1) \leqq O(M)-1,$$

i.e. the validity of Proposition 37 follows.

NOTE 2. Let $\mathfrak{R}_2$ be the (finite) projective plane over $\mathsf{F}$, consisting accordingly of $q^2+q+1$ points. We take a straight line $g$ in $\mathfrak{R}_2$ and denote its points by $P_0, \ldots, P_q$. Let $B(P_i)$ be the bundle of straight lines in $\mathfrak{R}_2$ with the carrier $P_i$ $(i=0, \ldots, q)$. After we have cancelled the point $P_i$ from the straight lines of $B(P_i)$ different from $g$, the remaining $q$ affine straight lines, each consisting of $q$ points, form a classification $C_i$ of the affine plane $\mathfrak{R}_2\setminus g$. $C_0, \ldots, C_q$ are pairwise almost disjoint so that Proposition 37 can be applied to them with

$$r = q+1, \quad M = \mathfrak{R}_2\setminus q, \quad k = q.$$

Thus the estimation

$$q+1 \leqq \frac{q^2-1}{q-1}$$

is obtained, using $O(M)=q^2$. Since the = sign holds here, we see from this example that Proposition 37 is "in general" strong.

We make the following remark in advance of a more exact investigation of the mappings (2), which begins here and leads, *inter alia*, to the proof of Theorem 19: The proposition that there are exactly $N$ permutations among the mappings (2) is of an existential nature, but it will, however, enable us to deduce a very precise general proposition. This will establish a connection with Problem II which we shall apply successfully.

For a function $F(\eta)\in \mathsf{F}$ $(\eta\in \mathsf{F})$ we write as $s_k(F(\eta))$ the $k$-th elementary symmetric function of all values $F(\eta)$ $(\eta\in\mathsf{F})$. (Consequently, $s_k(F(\eta))=s_k(F(\eta_1),\dots,F(\eta_q))$ holds with a notation introduced earlier, where $\mathsf{F}=\{\eta_1,\dots,\eta_q\}$.) The proposition that the mapping

$$\eta\to F(\eta)$$

is a permutation of $\mathsf{F}$ is (as has already been noted) equivalent to

$$\prod_{\eta\in\mathsf{F}}(x-F(\eta)) = x^q-x,$$

i.e. to

$$s_k(F(\eta)) = \begin{cases} 0 & (k=1,\dots,q-2,q),\\ -1 & (k=q-1).\end{cases}$$

It follows from this that there exist exactly $N$ elements $\xi\in\mathsf{F}$ such that

$$s_k(l_\eta(\xi)) = \begin{cases} 0 & (k=1,\dots,q-2,q),\\ -1 & (k=q-1).\end{cases}$$

For polynomials

$$F_\eta(x)\in\mathsf{F}[x] \qquad (\eta\in\mathsf{F})$$

(by analogy with $s_k(F(\eta))$) we write their $k$-th elementary symmetric polynomial as $s_k(F_\eta(x))$. The preceding result can then be given by saying that the system of equations

$$s_k(l_\eta(x)) = \begin{cases} 0 & (k=0,\dots,q-2,q),\\ -1 & (k=q-1)\end{cases} \tag{13}$$

(with the unknown $x$) has exactly $N$ solutions in $\mathsf{F}$.

This remains unchanged when (13) is subjected to a translation $x\to x+\xi$ $(\xi\in\mathsf{F})$. Therefore, more generally, the system

$$s_k(l_\eta(x+\xi)) = \begin{cases} 0 & (k=0,\dots,q-2,q),\\ -1 & (k=q-1)\end{cases} \tag{14}$$

has exactly $N$ solutions in $\mathsf{F}$ for all $\xi\in\mathsf{F}$.

(It might seem at first sight that this trivial generalization has given rise to nothing but complications and that it confers no advantage. We shall, however, very soon see it in a quite different light.)

It follows that those equations (14) whose left-hand side (relative to $x$) is of degree $<N$ must hold also "identically" (i.e. interpreted in $\mathsf{F}[x]$). We state this as

PROPOSITION 38. *Those equations* (14) *whose left-hand side is of degree* $<N$ *hold in* $\mathsf{F}[x]$ *for all* $\xi\in\mathsf{F}$.

Before proving this proposition, we shall settle the easy case $N=q$, in order to avoid being bothered by it later on.

The condition $N=q$ means that all the mappings (2) are permutations, i.e. that all the equations

$$f_\xi(x) = x^q - x \qquad (\xi\in\mathsf{F})$$

hold. We form the polynomial

$$f_x(y) = \prod_{\eta\in\mathsf{F}} (y - l_\eta(x))(\in \mathsf{F}[x, y])$$

with a second indeterminate $y$. (We use the rather unusual notation $f_x(y)$ to correspond with (9).) Then these equations can (after the substitution $x\rightarrow y$) be condensed into the single congruence

$$f_x(y) \equiv y^q - y \pmod{x^q - x}.$$

Since the total degree of the left-hand side is $q$, this congruence is equivalent to an equation

$$f_x(y) = y^q - y - \alpha(x^q - x),$$

where $\alpha$ can be an arbitrary element of $\mathsf{F}$. In view of $\alpha=\alpha^q$, the right-hand side is equal to

$$(y-\alpha x)^q - (y-\alpha x)(= \prod_{\eta\in\mathsf{F}} (y - \alpha x - \eta))$$

and therefore the last equation turns into

$$\prod_{\eta\in\mathsf{F}} (y - l_\eta(x)) = \prod_{\eta\in\mathsf{F}} (y - \alpha x - \eta).$$

Thus (after comparison of the factors) we obtain

THEOREM 20. *For the linear polynomials* (1), $N=q$ *holds* (*i.e. all the mappings* (2) *are permutations*) *if and only if for a suitable arrangement of them one has*

$$l_\eta(x) = \alpha x + \eta,$$

*where* $\alpha$ *may be any element of* $\mathsf{F}$.

For this reason, let us assume that $N \leqq q-1$ from now on.

As the left-hand side of (14) is of degree $\leqq k$ and it has the constant term $s_k(l_\eta(\xi))$, it follows from Proposition 38 that

$$s_k(l_\eta(\xi)) = 0 \qquad (k = 1, \ldots, N-1) \tag{15}$$

holds for all $\xi \in \mathsf{F}$. However, by (9) we have

$$f_\xi(x) = x^q - s_1(l_\eta(\xi))x^{q-1} + s_2(l_\eta(\xi))x^{q-2} - \ldots,$$

so that the $N-1$ terms after the term $x^q$ can be cancelled here. Thus we have proved

PROPOSITION 39. *If $N \leqq q-1$ the lower estimations*

$$f_\xi^\downarrow \geqq N \qquad (\xi \in \mathsf{F}) \tag{16}$$

*hold for the sinkings of the polynomials that accompany the mappings* (2).

The required general proposition consists of the estimations (16) (which are valid for all $\xi$). (It should be noted that in the settled case $N=q$ the estimations (16) do not hold, as then $f(x)$ is equal to $x^q - x$ for all $\xi \in \mathsf{F}$, whence it follows that $f_\xi^\downarrow = q-1 < q$.)

For a first application we shall settle the case $N = q-1$ completely. If $N = q-1$ there exists a (unique) $\varrho \in \mathsf{F}$ with $f_\varrho(x) \neq x^q - x$. On the other hand, Proposition 39 implies $f_\varrho^\downarrow \geqq q-1$, i.e.

$$f_\varrho(x) = x^q + \alpha x + \beta$$

holds for suitable $\alpha, \beta \in \mathsf{F}$. Since $f(x)$ is fully reducible, its zeros are also zeros of

$$(f_\varrho(x) - (x^q - x) =)(\alpha+1)x + \beta.$$

If $\alpha + 1 = 0$ we must have $\beta = 0$, i.e. $f_\varrho(x) = x^q - x$. As this is false, it follows that $f_\varrho(x)$ has only a single zero (which must, of course, be of multiplicity $q$). Thus

$$f_\varrho(x) = (x-\sigma)^q$$

holds for a suitable $\sigma \in \mathsf{F}$. By (9), it follows that

$$l_\eta(\varrho) = \sigma$$

for all $\eta \in \mathsf{F}$. This gives us that

$$l_\eta(x) = l_\eta(x) - l_\eta(\varrho) + \sigma,$$

so that, by (5), we obtain

$$l_\eta(x) = \alpha_\eta(x - \varrho) + \sigma \qquad (\eta \in \mathsf{F}).$$

As the mapping (2) is a permutation for all $\xi \in \mathsf{F} \setminus \varrho$, this holds in particular for $\xi = \varrho + 1$. Therefore

$$\eta \to \alpha_\eta + \sigma$$

is a permutation of F, whence the same follows for $\eta \to \alpha_\eta$. Accordingly, the polynomials (1) can be rearranged in such a way that $\alpha_\eta = \eta$ is true. Thus (after the change $\varrho \to \alpha$, $\sigma \to \beta$ of the notations) the "only if" part of the following theorem has been proved:

THEOREM 21. *For the linear polynomials* (1) $N = q-1$ *holds* (*i.e. the mappings* (2) *are permutations with a single exception*) *if and only if for a suitable arrangement*

$$l_\eta(x) = \eta(x-\alpha)+\beta \qquad (\eta \in \mathsf{F})$$

*holds, where* $\alpha$, $\beta$ *may be arbitrary elements of* F.

The "if" part, which has not yet been proved, is trivial.

For the case $N^* = N+1$ we can make Proposition 39 a little stronger. As already noted, this case means that $\eta \to l_\eta(\infty) = \alpha_\eta$ is a permutation of F. Owing to Theorem 20, this does not hold for $N = q$; therefore we have to consider here, also in what follows, only the case $N \leqq q-1$. Moreover, we also disregard the case $N = q-1$, which we know from Theorem 21, i.e. we here suppose that $N \leqq q-2$.

The desired strengthening will again be given by Proposition 38. We must bear in mind that the left-hand side of (14) is of degree $\leqq k$ and the coefficient of $x^k$ in it is (by (5)) the $k$-th elementary symmetric expression of the elements $\alpha_\eta$ $(\eta \in \mathsf{F})$. The latter vanishes if $N^* = N+1$ for the above reason, if $k \leqq q-2$. The same thing holds *a fortiori* for $k \leqq N$ (from $N \leqq q-2$). This means that now the left-hand side of (14) is for $k \leqq N$ of degree $\leqq k-1$, whence the validity of (15) follows by Proposition 38 for $k = 1, \ldots, N$. From this we obtain, as above:

PROPOSITION 40. *If* $N \leqq q-2$, $N^* = N+1$ *the estimation from below*

$$f_\xi^\downarrow \geqq N+1 \qquad (\xi \in \mathsf{F})$$

*holds for the sinking of the polynomials accompanying the mappings* (2).

(We shall later see the usefulness of this "slight" strengthening of Proposition 39.)

It has already been pointed out that all the $f_\xi(x)$ are fully reducible and that, for them,

$$f_\xi(x) \in \mathsf{F}[x], \quad f_\xi^\circ = q.$$

If, further, $\frac{q-1}{2} \leqq N \leqq q-1$ is satisfied, then

$$f_\xi^\downarrow \geqq \frac{q-1}{2}, \quad \text{i.e.} \quad f_\xi^{\circ\circ} \leqq \frac{q+1}{2} \qquad (\xi \in \mathsf{F})$$

follows from Proposition 38. This also holds for $N=q$, as the $f_\xi^\downarrow$ are then all equal to $q-1$. Thus we have proved

PROPOSITION 41. *In the case*

$$N \geqq \frac{q-1}{2} \tag{17}$$

*every* $f_\xi(x)$ $(\xi \in \mathsf{F})$ *is either a solution of Problem II or an element of* $\mathsf{F}[x^p]$.

Let us return to Proposition 39. If the "$>$" sign holds in (17), then $f_\xi^{\circ\circ} < \frac{q+1}{2}$ follows from it. Consequently the simplification now occurs in Proposition 41 that $f_\xi(x)$ can be a solution of Problem II only if it does not belong to the marginal case. In view of Theorem 17 $x^q - x$ is the only solution of this kind. We have thus derived

PROPOSITION 42. *If*

$$N > \frac{q-1}{2} \tag{18}$$

*every* $f_\xi(x)$ $(\xi \in \mathsf{F})$ *is either* $x^q - x$ *or an element of* $\mathsf{F}[x^p]$.

We now prove Theorem 19. If $N \leqq \frac{q-1}{2}$ or $N = q-1$ or $N=q$, then we have nothing to prove; we may therefore assume that

$$\frac{q-1}{2} < N < q-1. \tag{19}$$

We must now prove that $N$ then lies in an interval

$$\left(q - \frac{q-1}{p^e-1},\; q-1-\frac{q-1}{p^e+1}\right) \qquad \left(e = 1, \ldots, \left[\frac{n}{2}\right]\right). \tag{20}$$

Since $N$ is the number of permutations appearing in (2), i.e. the number of $\xi \in \mathsf{F}$ such that $f_\xi(x) = x^q - x$, it follows from Proposition 42 that there are exactly $q-N$ polynomials $f_\xi(x)$ $(\xi \in \mathsf{F})$ in $\mathsf{F}[x^p]$. The exponents of these $f_\xi(x)$ are $>1$. Let $p^e$ $(1 \leqq e \leqq n)$ be the least of these exponents. We show that necessarily $e \leqq \left[\frac{n}{2}\right]$ and that $N$ lies in the interval (20), whence follows Theorem 19.

First we show that

$$e < n. \tag{21}$$

If $e=n$, then there exists a $\varrho \in \mathsf{F}$ such that $f_\varrho(x) \in \mathsf{F}[x^{p^n}]$ $(=\mathsf{F}[x^q])$. This means that $f_\varrho(x) = (x-\sigma)^q$ for a suitable $\sigma \in \mathsf{F}$, and therefore

$$l_\eta(\varrho) = \sigma$$

follows from (9) for all $\eta \in \mathsf{F}$. Accordingly the kernel of the mapping $\eta \to l_\eta(\varrho)$ consists of a single class (equal to $\mathsf{F}$). Thus Proposition 36 tells us that the kernel of the mapping (2) is trivial for all the $\xi \in \mathsf{F} \setminus \varrho$, i.e. that these mappings are permutations. This means that $N = q-1$, but this is impossible, by (19). The contradiction that we have thus obtained proves (21).

We choose a $\xi \in \mathsf{F}$ with

$$f_\xi(x) \in \mathsf{F}[x^{p^e}] \setminus \mathsf{F}[x^{p^{e+1}}].$$

Hence it follows that there exists a $g(x)$ such that

$$f_\xi(x) = g(x)^{p^e}, \quad \underset{1}{g(x) \in} \mathsf{F}[x] \setminus \mathsf{F}[x^p], \quad g^\circ = \frac{q}{p^e}. \tag{22}$$

As, moreover, $f(x)$ is fully reducible, (22) implies the same for $g(x)$, so that

$$g^{\circ\circ} \geqq \frac{q+p^e}{p^e(p^e+1)}$$

must hold, by Theorem 18. By $(22_1)$ we have $f_\xi^{\circ\circ} = p^e g^{\circ\circ}$, too, whence

$$f_\xi^{\circ\circ} \geqq \frac{q+p^e}{p^e+1} \left(= 1 + \frac{q-1}{p^e+1}\right).$$

The left-hand side is equal to $q - f_\xi^{\downarrow}$, therefore it is, by Proposition 39, not greater than $q-N$; consequently the estimation from above

$$N \leqq q-1-\frac{q-1}{p^e+1} \tag{23}$$

holds.

In order to estimate $N$ from below, we consider the set $\mathfrak{M}$ of the $q-N$ elements $\xi \in \mathsf{F}$ for which

$$f_\xi(x) \in \mathsf{F}[x^p].$$

Because of the minimum property of $p^e$, these $f_\xi(x)$ lie in $\mathsf{F}[x^{p^e}]$. If therefore $C_\xi$ is the kernel of the mapping (2), it follows that for $\xi \in \mathfrak{M}$ the classes appearing in $C_\xi$ contain at least $p^e$ elements. As on account of Proposition 36 these $C_\xi$ are also pairwise almost disjoint, the estimation from above

$$q-N \leqq \frac{q-1}{p^e-1}$$

follows from Proposition 37. Consequently, the estimation from below

$$N \geqq q - \frac{p-1}{p^e - 1} \tag{24}$$

holds.

In view of (23) and (24) the inequality

$$\frac{q-1}{p^e-1} - 1 - \frac{q-1}{p^e+1} \geqq 0$$

must be correct. This means that

$$\frac{2(q-1)}{p^{2e}-1} \geqq 1,$$

consequently (because of $q = p^n$) $2e \leqq n$, i.e.

$$e \leqq \left[\frac{n}{2}\right],$$

as required above. Furthermore, since by (23) and (24) $N$ lies in the interval (20), Theorem 19 is proved.

Crudely, Theorem 19 asserts that the number $N$ of permutations occurring in (2) has only a few values in the interval $\left(\frac{q-1}{2}, q\right)$. We therefore raise the question of the determination of all linear systems (1) such that $N \geqq \frac{q-1}{2}$. A further problem is then to calculate the exact value of $N$ for each system (1) of this kind. (For $N = q$ and $N = q-1$ the answer is contained in Theorems 20 and 21.)

Although there is some hope of the solvability of these certainly very difficult problems, we have investigated them only in the case:

$$\boxed{q = p, \quad \mathsf{F} = \mathsf{F}_p, \quad n = 1.}$$

In the remainder of this section we shall restrict ourselves to this case and investigate it completely. We shall also assume that

$$p \neq 2,$$

as the case $p = 2$ is trivial.

The system (1) is now of the form

$$l_\eta(x) \in \mathsf{F}_p[x] \qquad (\eta \in \mathsf{F}_p, \quad l_\eta^\circ \leqq 1). \tag{25}$$

We call this system *polarized* if in a suitable arrangement of its elements they can be written in the form

$$\begin{aligned} &a(x) \\ &a(x)+\mu b(x) \qquad (\mu \in \mathsf{F}_p^+), \\ &a(x)+\nu c(x) \qquad (\nu \in \mathsf{F}_p^-), \end{aligned} \tag{26}$$

where $a(x)$, $b(x)$, $c(x)$ are suitable (linear) polynomials from $\mathsf{F}_p[x]$ and where we have put

$$\mathsf{F}_p^+ = \{\varkappa \in \mathsf{F}_p : \chi(\varkappa) = 1\}, \quad \mathsf{F}_p^- = \{\varkappa \in \mathsf{F}_p : \chi(\varkappa) = -1\}.$$

(Accordingly, $\mathsf{F}_p^+$ is the subgroup $\mathsf{F}_p^2 \setminus 0$ of $\mathsf{F}_p \setminus 0$ of index 2 and $\mathsf{F}_p^-$ is the class of $\mathsf{F}_p \setminus 0$ mod $\mathsf{F}_p^+$, which is different from $\mathsf{F}_p^+$. Of course, $\mathsf{F}_p^+$ could have been used instead of $\mathsf{F}_p^-$ in (26), if at the same time we replaced $c(x)$ by a suitable associated polynomial, but we prefer to retain (26) for expediency.)

PROPOSITION 43. *If for a system* (25) *of linear polynomials over* $\mathsf{F}_p$ *with* $p \neq 2$ *the number* $N$ *of permutations among the mappings*

$$\eta \to l_\eta(\xi) \qquad (\xi \in \mathsf{F}_p) \tag{27}$$

*of* $\mathsf{F}_p$ *(into itself) satisfies the condition*

$$N \geqq \frac{p-1}{2} \tag{28}$$

*then* (25) *is polarized.*

Although this proposition is quite profound, its proof, based on the foregoing results, will not be particularly difficult.

By Theorem 19 and (28), only the cases $N = \dfrac{p-1}{2}, p-1, p$ are possible; use of this fact would however not make the proof any shorter. We therefore admit all the possibilities (28) and disregard Theorem 19.

We shall begin the proof with a few reductions. First of all, if $p=3$, Proposition 43 is trivial, as any system of three linear polynomials from $\mathsf{F}_3[x]$ is obviously polarized. Let us therefore assume that

$$p \geqq 5.$$

We shall prove the Proposition for the easy cases $N = p-1$, $p$ straight away. By Theorems 21 and 20 we must show that the systems

$$l_\eta(x) = \eta(x-\alpha)+\beta \qquad (\eta \in \mathsf{F}_p;\ \alpha, \beta \in \mathsf{F}_p)$$

and

$$l_\eta(x) = \alpha x+\eta \qquad (\eta \in \mathsf{F}_p;\ \alpha \in \mathsf{F}_p)$$

are both polarized. (The elements $\alpha$, $\beta$ are fixed in both cases.) If we first choose

$$a(x) = \beta, \quad b(x) = c(x) = x-\alpha$$

and then take

$$a(x)=\alpha x, \quad b(x)=c(x)=1$$

in the representation (26) of the polarized systems, we obtain precisely the preceding systems and the assertion is thus proved.

Let us therefore assume that

$$\frac{p-1}{2} \leqq N \leqq p-2 \tag{29}$$

from now on.

We now introduce the following definitions: A system (25) is called *centred* if $l_0(x)=0$ (i.e. all the mappings (27) have the fixed element 0); it is said to be *i-normal* ($i=0, 1$) if the coefficients of $x^i$ in it are all the (different) elements of $\mathsf{F}_p$.

We next introduce the coefficient representation

$$l_\eta(x) = \alpha_\eta x + \beta_\eta \qquad (\eta \in \mathsf{F}_p) \tag{30}$$

of the system (25). Thus this system is centred, 0-normal, or 1-normal, according as $\alpha_0=\beta_0=0$, or $\eta \to \beta_\eta$ is a permutation of $\mathsf{F}_p$, or the same holds for $\eta \to \alpha_\eta$, respectively.

We now try to reduce the proof of Proposition 43 to the case of a centred 1-normal system (25); we shall achieve this in three steps.

Let $l(x) \in \mathsf{F}_p[x]$ be an arbitrary linear polynomial; then the number $N$ and the property of being polarized are invariant under the transformation

$$l_\eta(x) \to l_\eta(x) + l(x)$$

of system (25); for this transformation causes the mappings (27) to become

$$\eta \to l_\eta(\xi) + l(\xi) \qquad (\xi \in \mathsf{F}_p)$$

where $N$ is not influenced by the term $l(\xi)$, which is independent of $\eta$. And further, if the system (25) is polarized, i.e. of the form (26), then a similar statement holds with $a(x)+l(x)$ in place of $a(x)$. Thus both statements are correct.

It therefore suffices to prove Proposition 43 for this transformed system. Apply this with $l(x) = -l_0(x)$. Since $l_0(x)$ is then transformed into 0, we have shown that if suffices to prove Proposition 43 for centred systems (25). We shall therefore henceforth assume that

$$l_0(x)=0 \qquad (\text{i.e. } \alpha_0=\beta_0=0). \tag{31}$$

A translation $x \to x+\varkappa$ ($\varkappa \in \mathsf{F}_p$) turns (25) into the system

$$l_\eta(x+\varkappa) \qquad (\eta \in \mathsf{F}_p).$$

The number $N$ as well as the properties of being polarized and centred are obviously invariant under this translation. Since we now have $N>0$, it is possible to choose $\varkappa$ in such a way that

$$\eta \to l_\eta(\varkappa)$$

is a permutation of $\mathsf{F}_p$. As, however, we have

$$l_\eta(x+\varkappa) = \alpha_\eta x + l_\eta(\varkappa)$$

from (30), it follows from this that it is sufficient to prove Proposition 43 for centred 0-normal systems (25). We shall assume that we have such a system (25) from now on.

To carry out the last desired reduction, let us consider only those of our mappings (27) for which $\xi \neq 0$, i.e. the (cf. 30))

$$\eta \to l_\eta(\xi) = \alpha_\eta \xi + \beta_\eta \qquad (\xi \in \mathsf{F}_p \setminus 0).$$

(Thus the mapping $\eta \to l_\eta(0) = \beta_\eta$ has been disregarded here.) Since (25) is 0-normal, it follows that exactly $N-1$ permutations appear among these mappings. The same also holds for the mappings

$$\eta \to \frac{1}{\xi} l_\eta(\xi) = \beta_\eta \frac{1}{\xi} + \alpha_\eta,$$

which can be written (in another order) as

$$\eta \to \beta_\eta \xi + \alpha_\eta \qquad (\xi \in \mathsf{F}_p \setminus 0).$$

Accordingly, we introduce the linear polynomials

$$\bar{l}_\eta(x) = \beta_\eta x + \alpha_\eta \left( = x l_\eta \left( \frac{1}{x} \right) \right) \qquad (\eta \in \mathsf{F}_p) \tag{32}$$

and call (32) the *reciprocal system* to (25). Writing the number of permutations of $\mathsf{F}_p$ (into itself) that appear among the mappings

$$\eta \to \bar{l}_\eta(\xi) (= \beta_\eta \xi + \alpha_\eta) \qquad (\xi \in \mathsf{F}_p) \tag{33}$$

belonging to the system (32) as $\bar{N}$, we infer from the last statement that

$$N-1 \leqq \bar{N} \leqq N. \tag{34}$$

(More precisely, $\bar{N}=N$ or $\bar{N}=N-1$, according as $\eta\to\bar{l}_\eta(0)=\alpha_\eta$ is or is not a permutation, i.e. according as (33) is or not 0-normal.) Since (25) is centred and 0-normal, the reciprocal system (33) is centred and 1-normal.

We still have to go from (33) over to the extended mappings (cf. (8))

$$\eta\to\bar{l}_\eta(\xi) \qquad (\xi\in \mathsf{F}_p^*). \tag{35}$$

This means that the single mapping

$$\eta\to\bar{l}_\eta(\infty)=\beta_\eta \tag{36}$$

of $\mathsf{F}_p$ has been added to the mappings (33); this mapping is a permutation, according to our former remark. Let us write $\bar{N}^*$ for the number of the permutations appearing in (35). Then we have

$$\bar{N}^* = \bar{N}+1. \tag{37}$$

Now the system

$$xa\left(\frac{1}{x}\right),$$

$$xa\left(\frac{1}{x}\right)+\mu xb\left(\frac{1}{x}\right) \qquad (\mu\in\mathsf{F}_p^+),$$

$$xa\left(\frac{1}{x}\right)+\nu xc\left(\frac{1}{x}\right) \qquad (\nu\in\mathsf{F}_p^-),$$

that comes from forming the reciprocal of the polarized system (26), is likewise polarized (and vice versa); it is therefore sufficient to prove that system (35) is polarized. The intended (final) reduction in the proof of Proposition 43 consists in this fact. It is essential in this reduction that the system (35) is (not only centred but also) 1-normal, as has already been pointed out; this will be used successfully below.

From (29) and (34) we infer that $N\leqq p-2$. From this and (37),

$$g_\xi^! \geqq \bar{N}+1 \qquad (\xi\in\mathsf{F}_p) \tag{38}$$

follows from Proposition 40 (by application to the reciprocal system (32)), where

$$g_\xi(x) = \prod_{\eta\in\mathsf{F}_p}(x-\beta_\eta\xi-\alpha_\eta) \qquad (\xi\in\mathsf{F}_p) \tag{39}$$

is the polynomial accompanying the mapping (33). (Consequently we should have written $\bar{f}_\xi(x)$ in place of $g_\xi(x)$, but we have chosen a simpler notation in preference.)

Again from (29) and (34)

$$g_\xi^{!} \geqq \frac{p-1}{2} \qquad (\xi \in \mathsf{F}_p)$$

follows from (38). It follows (by $g_\xi^{\circ\circ} = p - g_\xi^{!}$) that each $g_\xi(x)\,(\xi \in \mathsf{F}_p)$ either is a solution (for $q=p$) of Problem II or lies in $\mathsf{F}_p[x^p]$. From Theorem 17 (see also Example 1, following the theorem) this means that each $g_\xi(x)$ must be one of the polynomials

$$\begin{gathered}(x+\varkappa)\left((x+\varkappa)^{\frac{p-1}{2}}-\sigma\right)\left((x+\varkappa)^{\frac{p-1}{2}}-\sigma\tau\right) \quad (\varkappa \in \mathsf{F}_p;\ \sigma = \pm 1;\ \tau = 0, 1);\\ x^p - x; \quad (x-\lambda)^p \qquad (\lambda \in \mathsf{F}_p).\end{gathered} \tag{40}$$

From this (from (39)) and from the fact that the system (32) is centred and 1-normal we shall be able to deduce that the latter is polarized, which will complete the proof of Proposition 43.

Since (32) is centred (i.e. $\alpha_0 = \beta_0 = 0$), the divisibility

$$x | g_\xi(x)$$

follows from (39) for $\xi \in \mathsf{F}_p$. For this reason only those of the polynomials (40) divisible by $x$ will henceforth be considered. For $(40_1)$ this divisibility is valid if and only if either $\varkappa = 0$ or $\varkappa \neq 0$ and $\sigma = \varkappa^{\frac{p-1}{2}}$ (here $\tau$ may be 0 or 1, arbitrarily); the same question for $(40_2)$ and $(40_3)$ is trivial. It follows that each $g_\xi(x)$ is one of the polynomials

$$x\left(x^{\frac{p-1}{2}}-\sigma\right)\left(x^{\frac{p-1}{2}}-\sigma\tau\right) \qquad (\sigma = \pm 1;\ \tau = 0, 1),$$

$$(x+\varkappa)\left((x+\varkappa)^{\frac{p-1}{2}}-\varkappa^{\frac{p-1}{2}}\right)\left((x-\varkappa)^{\frac{p-1}{2}}-\varkappa^{\frac{p-1}{2}}\tau\right) \qquad (\varkappa \in \mathsf{F}_p \setminus 0;\ \tau = 0, 1),$$

$$x^p - x,$$

$$x^p.$$

For pairwise disjoint sets $M_1, \ldots, M_r$ and natural numbers $k_1, \ldots, k_r$ we denote by

$$k_1 M_1 \cup \ldots \cup k_r M_r$$

the system consisting of the elements of $M_1 \cup \ldots \cup M_r$, where we count the elements of $M_i$ $k_i$ times ($i = 1, \ldots, r$). Using this abbreviation, the foregoing statement can be formulated as follows: The zeros of an accompanying polynomial $g_\xi(x)$, i.e. the images appearing as the result of a mapping (33) (counted with their multiplicities) are given for each $\xi \in \mathsf{F}_p$ by one of the following systems:

$$\mathfrak{S}_1 = \frac{p+1}{2} \cdot 0 \cup 1 \cdot \mathsf{F}_p^+,$$

$$\mathfrak{S}_2 = \frac{p+1}{2} \cdot 0 \cup 1 \cdot \mathsf{F}_p^-,$$

$$\mathfrak{S}_3 = 1 \cdot 0 \cup 2 \cdot \mathsf{F}_p^+,$$

$$\mathfrak{S}_4 = 1 \cdot 0 \cup 2 \cdot \mathsf{F}_p^-,$$

$$\mathfrak{S}_5 = \frac{p+1}{2} \cdot -\varkappa \cup 1 \cdot (-\varkappa + \varkappa \mathsf{F}_p^+) \qquad (\varkappa \in \mathsf{F}_p \setminus 0),$$

$$\mathfrak{S}_6 = 1 \cdot -\varkappa \cup 2 \cdot (-\varkappa + \varkappa \mathsf{F}_p^+) \qquad (\varkappa \in \mathsf{F}_p \setminus 0),$$

$$\mathfrak{S}_7 = 1 \cdot \mathsf{F}_p,$$

$$\mathfrak{S}_8 = p \cdot 0.$$

(Of course, $-\varkappa + \varkappa \mathsf{F}_p^+$ is to be thought of as the set of all $-\varkappa + \varkappa\omega$ such that $\omega \in \mathsf{F}_p^+$.) It must be emphasized that, apart from the last two, each of these systems contains exactly $\frac{p-1}{2}$ different elements of $\mathsf{F}_p \setminus 0$; the 0 appears in $\mathfrak{S}_5$ and $\mathfrak{S}_6$ with the multiplicities 1 and 2, respectively. We bear these points in our mind in what follows.

We distinguish three cases.

Case I: All the $\frac{\alpha_\eta}{\beta_\eta}$ $(\eta \in \mathsf{F}_p \setminus 0)$ are equal. (Note that the denominator here runs through all the elements of $\mathsf{F}_p \setminus 0$, it is consequently not equal to 0.) There thus exists a $\lambda \in \mathsf{F}_p$ with $\alpha_\eta = \lambda \beta_\eta$ $(\eta \in \mathsf{F}_p \setminus 0)$; this holds also for $\eta = 0$ as $\alpha_0 = \beta_0 = 0$. Thus the system (32) consists of the polynomials

$$\bar{l}_\eta(x) = \beta_\eta(x+\lambda) \qquad (\eta \in \mathsf{F}_p).$$

This system agrees (apart from the order of the elements) with the case

$$a(x) = 0, \quad b(x) = c(x) = x + \lambda$$

of (26) and it is consequently polarized.

Case II: All the $\frac{\alpha_\eta}{\beta_\eta}$ $(\eta \in \mathsf{F}_p \setminus 0)$ are different. Then there exists a (unique) $\lambda \in \mathsf{F}_p$ such that the values of these quotients make up the difference set $\mathsf{F}_p \setminus \lambda$;

the order of succession of the polynomials (32) can therefore be chosen in such a way that

$$\frac{\alpha_\eta}{\beta_\eta} = \eta + \lambda \qquad (\eta \in \mathsf{F}_p \setminus 0).$$

It follows (using $\alpha_0 = \beta_0 = 0$) that

$$\alpha_\eta = \beta_\eta(\eta + \lambda) \qquad (\eta \in \mathsf{F}_p).$$

After substituting into (32) we have

$$\bar{l}_\eta(x) = \beta_\eta(x + \eta + \lambda) \qquad (\eta \in \mathsf{F}_p)$$

and therefore the mappings (33) are of the form

$$\eta \to \bar{l}_\eta(\xi) = \beta_\eta(\xi + \eta + \lambda) \qquad (\xi \in \mathsf{F}_p).$$

When $\xi \neq -\lambda$ not only $\eta = 0$ but also $\eta = -\xi - \lambda$ ($\neq 0$) is mapped on 0; it follows that at most one permutation appears among these mappings. This means that $\bar{N} \leqq 1$ and it follows also that in the case $\bar{N} = 1$ the mapping (corresponding to the value $\xi = -\lambda$)

$$\eta \to \beta_\eta \eta$$

is a permutation.

If the latter holds, we obtain (using the fact that $\beta_0 = 0$ and that $\eta \to \beta_\eta$ is also a permutation)

$$\prod_{\eta \in \mathsf{F}_p \setminus 0} \beta_\eta \eta = \prod_{\eta \in \mathsf{F}_p \setminus 0} \beta_\eta,$$

whence after a reduction we get that

$$\prod_{\eta \in \mathsf{F}_p \setminus 0} \eta = 1.$$

This is a contradiction, as the left-hand side is equal to $-1$, by Wilson's theorem. As $\bar{N} = 1$ therefore impossible, $\bar{N} = 0$ must hold. Hence and from (34) $N \leqq 1$ follows. Since, however, $p \geqq 5$ and (28) imply that $N \geqq 2$, we have proved that the case in question is impossible.

Case III: The $\frac{\alpha_\eta}{\beta_\eta}$ $(\eta \in \mathsf{F}_p \setminus 0)$ are neither (all) equal nor (all) different. We choose two different elements $\eta_1, \eta_2 \in \mathsf{F}_p \setminus 0$ such that

$$\frac{\alpha_{\eta_1}}{\beta_{\eta_1}} = \frac{\alpha_{\eta_2}}{\beta_{\eta_2}} = -\lambda$$

for a suitable $\lambda \in \mathsf{F}_p$. Hence

$$\beta_{\eta_1}\lambda + \alpha_{\eta_1} = \beta_{\eta_2}\lambda + \alpha_{\eta_2} = 0,$$

where the $\beta_\eta\lambda+\alpha_\eta$ $(\eta\in\mathsf{F}_p)$ are not all equal to zero. We may assume that $\lambda=0$, from the fact that we may pass from (32) over to the polynomials

$$l_\eta(x+\lambda) = \beta_\eta x+(\beta_\eta\lambda+\alpha_\eta) \qquad (\eta\in\mathsf{F}_p),$$

whereby (32) does not lose its property of being centred and 1-normal. Then, however, we have simply

$$\alpha_{\eta_1}=\alpha_{\eta_2}=0,$$

while not all $\alpha_\eta$ $(\eta\in\mathsf{F}_p)$ are zero. Hence and from $\alpha_0=0$ we conclude that the system of all the

$$\bar{l}_\eta(0)=\alpha_\eta \qquad (\eta\in\mathsf{F}_p)$$

agrees with $\mathfrak{S}_1$ or $\mathfrak{S}_2$. This means that, for a suitable arrangement of the polynomials (32),

$$\alpha_\eta=\gamma\eta \qquad (\eta\in\mathsf{F}_p^+),$$
$$\alpha_\eta=0 \qquad (\eta\in\mathsf{F}_p^-)$$

holds for some $\gamma\in\mathsf{F}_p\setminus 0$. Then we have, again by (32) $(\bar{l}_0(x)=0$ and$)$

$$\bar{l}_\eta(x) = \beta_\eta x+\gamma\eta \quad (\eta\in\mathsf{F}_p^+), \qquad \bar{l}_\eta(x) = \beta_\eta x \qquad (\eta\in\mathsf{F}_p^-). \tag{41}$$

(We may arbitrarily rearrange the second half $(41_2)$ of this system.)

Let us put

$$\xi_\omega = -\frac{\gamma\omega}{\beta_\omega}(\neq 0) \qquad (\omega\in\mathsf{F}_p^+). \tag{42}$$

$(41_1)$ and (42) imply that

$$\bar{l}_\omega(\xi_\omega)=0 \qquad (\omega\in\mathsf{F}_p^+). \tag{43}$$

Let us consider the system

$$\bar{l}_\eta(\xi_\omega) \qquad (\eta\in\mathsf{F}_p) \tag{44}$$

for an arbitrary $\omega\in\mathsf{F}_p^+$. (The remaining $\omega$ will not be used.) From (43) and $l(x)=0$, the 0 appears at least twice in the latter system; it further follows from $(41_2)$ that the $\dfrac{p-1}{2}$ elements

$$\bar{l}_\eta(\xi_\omega)(=\beta_\eta\xi_\omega) \qquad (\eta\in\mathsf{F}_p^-)$$

from (44) are different from one another and from 0. This clearly means that the system (44) must agree with $\mathfrak{S}_1$ or $\mathfrak{S}_2$ or with one of the $\mathfrak{S}_6$. We distinguish two cases.

Let (44) be a $\mathfrak{S}_1$ or a $\mathfrak{S}_2$. This clearly means that

$$\bar{l}_\eta(\xi_\omega) = \beta_\eta \xi_\omega + \gamma\eta = 0 \qquad (\eta \in \mathsf{F}_p^+)$$

and

$$\bar{l}_\eta(\xi_\omega) = \beta_\eta \xi_\omega = \delta_\eta \qquad (\eta \in \mathsf{F}_p^-)$$

for a $\delta \in \mathsf{F}_p \setminus 0$, after the second "half system" ($41_2$) has been suitably rearranged. (In the first case $\delta$ must be taken from $\mathsf{F}_p^-$, but in the second from $\mathsf{F}_p^+$. Otherwise it may be chosen arbitrarily in both cases.) The $\beta_\eta$ ($\eta \neq 0$) can be calculated from these equations. After substitution of their values determined in this way the system (41) takes the form

$$\bar{l}_\eta(x) = \begin{cases} \eta\left(-\dfrac{\gamma}{\xi_\omega}x + \gamma\right) & (\eta \in \mathsf{F}_p^+), \\[2ex] \eta\dfrac{\delta}{\xi_\omega}x & (\eta \in \mathsf{F}_p^-). \end{cases}$$

Accordingly (and as $l_0(x) = 0$) the system (32) is polarized.

In the remaining case we shall derive a contradiction. The system (44) is now one of the systems $\mathfrak{S}_6$ (for a suitable $\varkappa \in \mathsf{F}_p \setminus 0$). It follows that each element (44), different from 0, must be one of the elements

$$-\varkappa; \quad -\varkappa + \varkappa\eta \qquad (\eta \in \mathsf{F}_p^+ \setminus 1).$$

On the other hand, as we have already mentioned by ($41_2$), the elements

$$\beta_\eta \xi_\omega \qquad (\eta \in \mathsf{F}_p^-)$$

appear among the elements (44) and they are different from one another and from 0; for this reason these elements must agree (to within a rearrangement) with the previous ones. The latter can be written with the aid of an arbitrary $\nu \in \mathsf{F}_p^-$ in the form

$$\beta_{\nu\eta} \xi_\omega \qquad (\eta \in \mathsf{F}_p^+).$$

It follows therefore that after a suitable rearrangement of the half system ($41_2$) the equations

$$\beta_\nu \xi_\omega = -\varkappa,$$

$$\beta_{\nu\eta} \xi_\omega = -\varkappa + \varkappa\eta \qquad (\eta \in \mathsf{F}_p^+ \setminus 1)$$

hold. Hence

$$\beta_\nu = -\frac{\varkappa}{\xi_\omega},$$

$$\beta_{\nu\eta} = \varkappa\frac{\eta - 1}{\xi_\omega} \qquad (\eta \in \mathsf{F}_p^+ \setminus 1),$$

so that the system (41) turns into

$$\begin{aligned} \bar{l}_\eta(x) &= \beta_\eta x + \gamma\eta \qquad (\eta \in \mathsf{F}_p^+); \\ \bar{l}_\nu(x) &= -\frac{\varkappa}{\xi_\omega} x; \quad \bar{l}_{\nu\eta}(x) = \varkappa \frac{\eta}{\xi_\omega} x \qquad (\eta \in \mathsf{F}_p^+ \setminus 1). \end{aligned} \tag{45}$$

We choose two different elements $\eta_1, \eta_2 \in \mathsf{F}_p^+$ and determine $\xi_0 \in \mathsf{F}_p$ by

$$\bar{l}_{\eta_1}(\xi_0) = \bar{l}_{\eta_2}(\xi_0), \tag{46}$$

i.e.

$$\beta\eta_1\xi_0 + \gamma\eta_1 = \beta\eta_2\xi_0 + \gamma\eta_2. \tag{47}$$

Since $\eta_1 \neq \eta_2$ and $\beta\eta_1 \neq \beta\eta_2$, $\xi_0$ is defined uniquely and, as $\gamma \neq 0$, is not equal to 0. We now consider the system

$$\bar{l}_\eta(\xi_0) \qquad (\eta \in \mathsf{F}_p); \tag{48}$$

this must be equal to one of the systems $\mathfrak{S}_1, \ldots, \mathfrak{S}_8$. We shall use it to deduce the desired contradiction.

By $(41_2)$ and $\xi_0 \neq 0$, the $\dfrac{p-1}{2}$ elements

$$\bar{l}_\eta(\xi_0) \qquad (\eta \in \mathsf{F}_p^-) \tag{49}$$

are different from one another and from 0. Therefore and by (46) the system (48) is different from $\mathfrak{S}_7$ and from $\mathfrak{S}_8$. It also clearly follows that each element of (48) either is 0 or belongs to the partial system (49). Let $\tau$ denote the left-hand side of (46). If $\tau \neq 0$, then (using (46)) it follows that $\tau$ occurs in the system (48) at least three times. As this is impossible, we must have $\tau = 0$. It follows again from (46) (by $\bar{l}_0(x) = 0$) that 0 occurs at least three times in (49), so that (48) must be either $\mathfrak{S}_1$ or $\mathfrak{S}_2$. It follows further that the set of elements (49) is equal to $\mathsf{F}_p^+$ or $\mathsf{F}_p^-$. From (45), however, (49) consists of the elements

$$-\frac{\varkappa}{\xi_\omega}\xi_0; \quad \varkappa\frac{\eta-1}{\xi_\omega}\xi_0 \qquad (\eta \in \mathsf{F}_p^+ \setminus 1).$$

After cancelling the common factors $\dfrac{\varkappa}{\xi_\omega}\xi_0$ we see that the set of the elements

$$-1; \quad \eta - 1 \qquad (\eta \in \mathsf{F}_p^+ \setminus 1)$$

is equal to $\mathsf{F}_p^+$ or $\mathsf{F}_p^-$. Thus all the character values

$$\chi(\eta - 1) \qquad (\eta \in \mathsf{F}_p^+ \setminus 1)$$

must be equal to $\chi(-1)$. This means that all elements

$$1 - \eta \qquad (\eta \in \mathsf{F}_p^+ \setminus 1)$$

belong to $F_p^+$. So,

$$\prod_{\eta \in F_p^+ \setminus 1} (x+\eta-1) \mid x^{\frac{p-1}{2}}-1.$$

Calculation gives the left-hand side the value

$$\frac{1}{x} \prod_{\eta \in F_p^+} (x+\eta-1) = \frac{1}{x}(-1)^{\frac{p-1}{2}} \prod_{\eta \in F_p^+} (1-x-\eta) = \frac{1}{x}\left((x-1)^{\frac{p-1}{2}}-(-1)^{\frac{p-1}{2}}\right).$$

Thus

$$(x-1)^{\frac{p-1}{2}}-(-1)^{\frac{p-1}{2}} \,\Big|\, x\left(x^{\frac{p-1}{2}}-1\right),$$

whence

$$x\left(x^{\frac{p-1}{2}}-1\right) = (x+\lambda)\left((x-1)^{\frac{p-1}{2}}-(-1)^{\frac{p-1}{2}}\right)$$

for a suitable $\lambda \in F_p$. Comparison of the coefficients of $x^{\frac{p-1}{2}}$ and $x^1$ gives

$$0 = \lambda - \frac{p-1}{2}, \quad -1 = \lambda(-1)^{\frac{p-3}{2}} \frac{p-1}{2},$$

i.e.

$$\lambda = -\frac{1}{2}, \quad -4 = (-1)^{\frac{p-3}{2}}$$

(these equations are to be interpreted as equations in $F_p$). The last equation, however, is impossible as $p \geqq 5$ (actually for all prime numbers $p$). This contradiction proves Proposition 43.

To draw further conclusions from this, let us consider, as a prelude to what follows, a polarized system (25) with the aim of trying to calculate the value of $N$ for it.

We return to the accompanying polynomials

$$f_\xi(x) = \prod_{\eta \in F_p} (x - l_\eta(\xi)) \qquad (\xi \in F_p).$$

Then $N$ is equal to the number of the $\xi$ such that $f_\xi(x) = x^p - x$.

To calculate $f_\xi(x)$, we write the given (polarized) system (25) in the form (26). As, however, $N$ is independent of $a(x)$, we may confine ourselves to the case $a(x)=0$. We clearly then have

$$f_\xi(x) = x \prod_{\mu \in F_p^+} (x-\mu b(\xi)) \prod_{\nu \in F_p^-} (x-\nu c(\xi)),$$

i.e.

$$f_\xi(x) = x\left(x^{\frac{p-1}{2}} - b(\xi)^{\frac{p-1}{2}}\right)\left(x^{\frac{p-1}{2}} + c(\xi)^{\frac{p-1}{2}}\right).$$

Thus

$$f_\xi(x) = x\left(x^{\frac{p-1}{2}} - \chi(b(\xi))\right)\left(x^{\frac{p-1}{2}} + \chi(c(\xi))\right).$$

This is equal to $x^p - x$ if and only if

$$\chi(b(\xi)) = \chi(c(\xi)) \neq 0,$$

which we may replace by

$$\chi(b(\xi)c(\xi)) = 1.$$

Writing the set of the zeros of $b(x)$ and $c(x)$ lying in $\mathsf{F}_p$ as $\mathfrak{N}$, on the strength of this result we may write down the formula

$$N = \sum_{\xi \in \mathsf{F}_p/\mathfrak{N}} \frac{1 - \chi(b(\xi)c(\xi))}{2},$$

i.e.

$$N = \frac{p - O(\mathfrak{N})}{2} + \frac{1}{2} \sum_{\xi \in \mathsf{F}_p} \chi(b(\xi)c(\xi)). \tag{50}$$

Now we have, thanks to Jacobsthal,

PROPOSITION 44. *For every $p \neq 2$ and $\varrho \in \mathsf{F}_p \setminus 0$*

$$\sum_{\xi \in \mathsf{F}_p} \chi(\xi^2 + \varrho) = -1. \tag{51}$$

To get a quick proof, let us write the left-hand side of (51) as $S$. Then

$$S \equiv \sum_{\xi \in \mathsf{F}_p} (\xi^2 + \varrho)^{\frac{p-1}{2}} \pmod p,$$

so that, after expanding and summing, we get

$$S \equiv -1 \pmod p.$$

As, moreover, trivially $|S| \leqq p$ and $2 \nmid S$, we see that $S = -1$, i.e. (51) follows.

More generally, the following statement also holds:

PROPOSITION 45. *For arbitrary $p \neq 2$ and $\varkappa, \lambda, \mu \in \mathsf{F}_p$*

$$\sum_{\xi \in \mathsf{F}_p} \chi(\varkappa^2 + \lambda\xi + \mu) = \begin{cases} -\chi(\varkappa) & (\lambda^2 - 4\varkappa\mu \neq 0), \\ (p-1)\chi(\varkappa) & (\varkappa \neq 0,\ \lambda^2 - 4\varkappa\lambda = 0). \\ p\chi(\mu) & (\varkappa = \lambda = 0). \end{cases} \tag{52}$$

This follows for $\varkappa = 0$ immediately and for $\varkappa \neq 0$ from Proposition 44.

We now go on to prove

THEOREM 22. *When $p \neq 2$ for a system of linear polynomials*

$$l_\eta(x) \in \mathsf{F}_p[x] \qquad (\eta \in \mathsf{F}_p,\ l_\eta^\circ \leqq 1) \tag{53}$$

*the number $N$ of permutations that appear among the mappings*

$$\eta \to l_\eta(\xi) \qquad (\xi \in \mathsf{F}_p) \tag{54}$$

*of* $\mathsf{F}_p$ *(into itself), is equal to at least* $\frac{p-1}{2}$ *if and only if the polynomials* (53) *are (apart from a possible rearrangement) of the form*

$$\begin{cases} a(x), \\ a(x)+\varrho(\alpha x+\beta) & (\varrho \in \mathsf{F}_p^+), \\ a(x)+\sigma(\gamma x+\delta) & (\sigma \in \mathsf{F}_p^-), \end{cases} \tag{55}$$

*where*

$$a(x) \in \mathsf{F}_p[x];\ a^\circ \leqq 1;\ \alpha, \beta, \gamma, \delta \in \mathsf{F}_p$$

*and one of the three cases*

1) $\alpha\delta-\beta\gamma \neq 0, \qquad \chi(\alpha\gamma) \leqq 0,$

2) $\alpha\delta-\beta\gamma = 0, \qquad \chi(\alpha\gamma)=1,$

3) $\alpha=\gamma=0, \qquad \chi(\beta\gamma)=1$

*holds. Corresponding to these three cases we have*

$$N=\frac{p-1}{2}, \quad p-1 \quad \textit{or} \quad p, \tag{56}$$

*respectively.*

Theorem 22 is given (with a fallacious proof) by Rédei [6, 12]. (See also Rédei [13].)

The proof is based on Proposition 43. We assume that

$$N \geqq \frac{p-1}{2}, \tag{57}$$

whence it follows that (53) is polarized, i.e. that it coincides with a system (55). We shall therefore also assume the latter to be the case. What we shall now do is therefore to determine the possible $\alpha$, $\beta$, $\gamma$, $\delta$ (under both assumptions) and calculate the exact value of $N$ in each case. (The polynomial $a(x)$ remains arbitrary, as $N$ is independent of it.)

We use formula (50). Writing the sum appearing in this formula as $S$, it can be put in the form

$$N=\frac{p-O(\mathfrak{N})}{2}+\frac{1}{2}S, \tag{58}$$

where (after applying (55))

$$S\left(=\sum_{\xi\in\mathsf{F}_p}\chi\big((\alpha x+\beta)(\gamma\xi+\delta)\big)\right)=\sum_{\xi\in\mathsf{F}_p}\chi\big(\alpha\gamma\xi^2+(\alpha\delta+\beta\gamma)\xi+\beta\delta\big) \tag{59}$$

holds and $\mathfrak{N}$ is the set of zeros of

$$(\alpha x+\beta)(\gamma x+\delta), \tag{60}$$

that lie in $\mathsf{F}_p$.

The condition (57) is, by (58), equivalent to

$$O(\mathfrak{N}) \leqq 1+S. \tag{61}$$

Now $S$ can be calculated with the aid of Proposition 45. For this purpose we put

$$\varkappa=\alpha\gamma,\quad \lambda=\alpha\delta+\beta\gamma,\quad \mu=\beta\delta,$$

whence

$$\lambda^2-4\varkappa\mu=(\alpha\delta-\beta\gamma)^2$$

so, by (52) and (59)

$$S=\begin{cases}-\chi(\alpha\gamma) & (\alpha\delta-\beta\gamma\neq 0),\\ (p-1)\chi(\alpha\gamma) & (\alpha\gamma\neq 0,\ \alpha\delta-\beta\gamma=0),\\ p\chi(\beta\delta) & (\alpha\gamma=\alpha\delta-\beta\gamma=0).\end{cases} \tag{62}$$

We distinguish several cases.

If $\alpha\delta-\beta\gamma\neq 0$, then clearly $O(\mathfrak{N})=1+\chi(\alpha\gamma)^2$. On the other hand, $S=-\chi(\alpha\gamma)$ holds by (62). The condition (57), i.e. (61), therefore turns into

$$\chi(\alpha\gamma)^2 \leqq -\chi(\alpha\gamma).$$

This is equivalent to

$$\chi(\alpha\gamma)\leqq 0.$$

We now have Case 1). It follows from (58) that

$$N=\frac{p-1-\chi(\alpha\gamma)^2-\chi(\alpha\gamma)}{2}=\frac{p-1}{2},$$

as stated above.

If $\alpha\delta-\beta\gamma=0$ and $\alpha\gamma\neq 0$, then $O(\mathfrak{N})=1$ and, moreover, it follows from (62) that $S=(p-1)\chi(\alpha\gamma)$, so that condition (61) becomes $\chi(\alpha\gamma)\geqq 0$. As $\alpha\gamma\neq 0$, the latter is equivalent to

$$\chi(\alpha\gamma)=1.$$

So we have come up against Case 2), and

$$N=\frac{p-1+p-1}{2}=p-1$$

follows, from (58), as stated.

Finally, if $\alpha\delta - \beta\gamma = 0$ and $\alpha\gamma = 0$ (and therefore also $\alpha\delta = \beta\gamma = 0$), then obviously

$$O(\mathfrak{N}) = p(1 - \chi(\beta\delta)^2)$$

and by (62)

$$S = p\chi(\beta\delta),$$

so that condition (61) becomes

$$p(1 - \chi(\beta\delta)^2) \leqq 1 + p\chi(\beta\delta)$$

which is equivalent to

$$\chi(\beta\delta) = 1.$$

Since this implies that ($\beta \neq 0, \delta \neq 0$) $\alpha = \gamma = 0$, Case 3) applies and, what is more, $O(\mathfrak{N}) = 0$, $S = p$; thus $N = p$, by (58). This completes the proof of Theorem 22.

(It is worth pointing out that part of Theorem 22 follows from Theorems 20 and 21; this cannot, however, be used to shorten the proof to any worthwhile extent.)

## § 35. Application to Hajós's theory

By the *schlicht product* $A_1 \ldots A_k$ of complexes $A_1, \ldots, A_k$ of an abelian group we understand the multiplicative analogue of the schlicht sum complexes of a module (defined in § 27); we write the schlicht product as $A_1 \bullet \ldots \ldots \bullet A_k$. The fundamental problem of Hajós's theory consists in the investigation of the schlicht product decompositions

$$G = A_1 \bullet \ldots \bullet A_k \tag{1}$$

of a finite abelian group $G$. It should be noticed that (1) is accordingly equivalent to the validity of $G = A_1 \ldots A_k$ and

$$O(G) = O(A_1) \ldots O(A_k).$$

We call decomposition (1) *normed* if $1 \in A_1, \ldots, A_k$. It is clear that it will be sufficient to investigate the normed schlicht decompositions.

As an application of Theorems 19, 20, 21, we prove

THEOREM 23. *In every normed schlicht product decomposition of an elementary abelian group, the order of which is equal to the square of a prime number, at least one factor must be a group.*

Before proceeding to the proof, we must point out that this seemingly modest theorem is very important in two respects. On the one hand, it has resisted every attempt at a direct proof and on the other, it serves as a steppingstone for a far-reaching generalization (see Rédei [8]).

Let $G$ be the given group with $O(G)=p^2$ and with the basis $\alpha$, $\beta$. It suffices to consider a normed decomposition of the form

$$G=A\bullet B \qquad (A, B=G;\ O(A)=O(B)=p). \tag{2}$$

We must prove that $A$ or $B$ is a group. In the case $p=2$ this is true and we therefore assume that $p\neq 2$.

Let $\varrho$ be a complex root of unity of order $p$. For each $z=0, \ldots, p-1$ a character $\omega_z$ of $G$ is then defined by

$$\omega_z(\alpha^i\beta^j)=\varrho^{iz+j} \qquad (i, j\in\mathscr{Z}).$$

It follows from (2) that

$$\sum_{\gamma\in G}\omega_z(\gamma) = \sum_{\varkappa\in A}\omega_z(\varkappa)\sum_{\lambda\in B}\omega_z(\lambda).$$

As the left-hand side vanishes, for all $z$ at least one factor of the right-hand side must vanish. Thus at least one factor of the right-hand side must vanish for at least $\frac{p+1}{2}$ values of $z$. We may assume that this is so for the first factor.

We now use the basis representation of the elements of $A$. This means that $A$ appears in the form

$$A=\{\alpha^{a_0}\beta^{b_0}, \ldots, \alpha^{a_{p-1}}\beta^{b_{p-1}}\} \qquad (a_0=b_0=0). \tag{3}$$

Then the above means that

$$\sum_{i=0}^{p-1}\varrho^{a_iz_i+b_i}=0$$

for at least $\frac{p+1}{2}$ values of $z$, i.e.

$$a_iz+b_i \qquad (i=0, \ldots, p-1)$$

is a representative system mod $p$.

From now on we again take the elements of $z$ to be elements of $\mathsf{F}_p$. Correspondingly, we interpret

$$l_i(x) = a_ix+b_i \qquad (i=0, \ldots, p-1) \tag{4}$$

as (linear) polynomials from $\mathsf{F}_p[x]$. Then the last statement asserts that among the $p$ mappings

$$i\to l_i(z) \qquad (z=0, \ldots, p-1)$$

of $\mathsf{F}_p$ into itself there are at least $\frac{p+1}{2}$ permutations. It follows from this by Theorem 19 that the number of these permutations must be equal to $p$

or $p-1$. Moreover, we may infer from this, by Theorems 20 and 21, that the system (4) agrees with a system

$$ax+i \qquad (i=0, \ldots, p-1)$$

or

$$i(x-a)+b(=ix-ia+b) \qquad (i=0, \ldots, p-1),$$

where $a$ and $b$ are fixed elements from $\mathscr{Z}$.

As $l_0(x)=0$, $a=0$ must hold in the first case and $b=0$ in the second. Comparison with (3) and (4) gives
$A=\{1, \beta, \ldots, \beta^{p-1}\}$ or, respectively, $A=\{1, \alpha\beta^{-a}, \alpha^2\beta^{-2a}, \ldots, \alpha^{p-1}\beta^{a-(p-1)a}\}$. We thus have that $A$ is the subgroup of $G$ generated by $\beta$ or by $\alpha\beta^{-a}$. This proves Theorem 23.

(It should be noted that the last part of the proof, based upon Theorems 19, 20 and 21, could just as well be based on Theorem 22. The proof thus obtained would not be essentially different, as Theorem 22 also depends on Theorem 19 while Theorems 20 and 21 are completely elementary. Theorem 19 seems to us to be crucial to the proof of Theorem 23. We reiterate that Theorem 19, on the other hand, stems from Theorem 17. Of course, only the prime case $q=p$ of Theorems 17 and 19 was required for the proof of Theorem 23, and, moreover, the part of Theorem 17 relating to the marginal case $f^{\circ\circ}=\dfrac{p+1}{2}$ and similarly the part of Theorem 19 relating to the "marginal case" $N=\dfrac{p-1}{2}$ remained unused.)

NOTE. In a recent work by E. Wittmann [15] the proof of Theorem 23 is given in the shortest possible from. This proof is, however, also based on the case $q=p$, $f^{\circ\circ}<\dfrac{p+1}{2}$ of Theorem 17.

## § 36. On the difference quotient of functions in finite fields

Let $f(\xi)$ be an arbitrary function in $\mathsf{F}$, that is the variable $\xi$ runs through all elements of $\mathsf{F}$ and the values $f(\xi)$ lie in $\mathsf{F}$. Starting from $f(\xi)$, let us use a second variable $\eta\in\mathsf{F}$ to form the *difference quotient* of $f(\xi)$:

$$\mathsf{F}(\xi, \eta)=\frac{f(\xi)-f(\eta)}{\xi-\eta}(=F(\eta, \xi)) \qquad (\xi\neq\eta). \tag{1}$$

Note 1. One could make (1) into a function from $\mathsf{F}\times\mathsf{F}$ to $\mathsf{F}^*(=\mathsf{F}\cup\infty)$, for instance by putting $\mathsf{F}(\xi, \xi)=\infty$. Then (1) would deal with the finite values of this function. We shall refrain from making this extension, but we may, however, treat (1), somewhat inaccurately, as a function of the two variables $\xi$, $\eta$ (though only for $\xi \neq \eta$).

NOTE 2. Of course, all the possible values for (1) are already included if we start from polynomials $f(x) \in \mathsf{F}[x]$. As then

$$\frac{f(x)-f(y)}{x-y} \in \mathsf{F}[x, y]$$

is also a polynomial, (1) can be obtained as the substitution value of this polynomial, belonging to the place $x=\xi$, $y=\eta$. The case $\xi=\eta$ is not an exception here and leads to the substitution value $f'(\xi)$ which is, however, (not class-invariant mod $x^q-x$, i.e.) not uniquely determined by the function $f(\xi)$, as the assumption

$$g(x) \equiv f(x) \qquad (\mathrm{mod}\ x^q - x) \quad \big(f(x),\ g(x) \in \mathsf{F}[x]\big)$$

implies merely the existence of a $h(x) \in \mathsf{F}[x]$ with

$$g(x) = f(x) + (x^q - x)h(x)$$

and beyond this the congruence

$$g'(x) \equiv f'(x) - h(x) \qquad (\mathrm{mod}\ x^q - x).$$

Therefore $g'(x) \equiv f'(x)$ (mod $x^q - x$) need not be fulfilled. (Although we could impose an unambiguous definition of $\mathsf{F}(\xi, \xi)$ in this way, for instance by the restriction $f^\circ \leqq q-1$, we reject this, as it would introduce an extraneous and useless complication into our discussions. So we stick to the original definition (1).)

We write the set of all difference quotients (1) as $\mathfrak{W} = \mathfrak{W}_f$, call this the *range of the difference quotient* of $f(\xi)$ and propose to investigate its cardinality $O(\mathfrak{W})$. We must initially be content to state that $1 \leqq O(\mathfrak{W}) \leqq q$, but it will later transpire that $O(\mathfrak{W})$ can assume only a "few" values in the interval $\left(1, \frac{q}{2}\right)$.

Fortunately, an approach to the set $\mathfrak{W}$ can be made by considering the difference set $\mathsf{F} \setminus \mathfrak{W}$. For, in order that a $\zeta \in \mathsf{F}$ should lie in $\mathsf{F}/\mathfrak{W}$ it is necessary and sufficient that for $\xi, \eta \in \mathsf{F}$, $\xi \neq \eta$, it always holds that

$$\frac{f(\xi)-f(\eta)}{\xi-\eta} \neq \zeta,$$

i.e. that

$$\eta\zeta - f(\eta) \neq \xi\zeta - f(\xi).$$

But this condition is equivalent to the fact that the mapping

$$\eta \to \eta\zeta - f(\eta)$$

of F (into itself) is a permutation. So, by a logical inversion we have proved

PROPOSITION 46. *For a function* $f(\xi) \in \mathsf{F}(\xi \in \mathsf{F})$ *a* $\zeta \in \mathsf{F}$ *belongs to the range* $\mathfrak{W}$ *of the difference quotient* (1) *of* $f(\xi)$ *if and only if the mapping*

$$\eta \to \eta\zeta - f(\eta) \qquad (\eta \in \mathsf{F}) \tag{2}$$

*of* F *(into itself) is not a permutation.*

If from the function $f(\xi)$ we now form the system of linear polynomials

$$\eta x - f(\eta) \qquad (\eta \in \mathsf{F}), \tag{3}$$

to which the family

$$\xi \to \eta\xi - f(\eta) \qquad (\xi \in \mathsf{F}) \tag{4}$$

of linear mappings of F (into itself) consequently belongs, then (after the substitution $\zeta \to \xi$) the mappings (2) are exactly the family dual to (4). For this reason the relation contained in Proposition 46, will be briefly called a *reciprocity* (between difference quotients and linear mappings in F). As a consequence of this reciprocity the results of § 34 lead us to statements in connection with the range $\mathfrak{W}$, as follows.

First of all, Proposition 46 gives us that

$$O(\mathfrak{W}) + N = q \tag{5}$$

always holds, where $N$ is the number of permutations among the mappings (2).

Thus it follows from Theorem 19 that $O(\mathfrak{W})$ always lies in one of the intervals

$$(0, 1); \quad \left(1 + \frac{q-1}{p^e+1}, \frac{q-1}{p^e-1}\right)\left(e = 1, \ldots, \left[\frac{n}{2}\right]\right); \quad \left(\frac{q+1}{2}, q\right).$$

The case $O(\mathfrak{W}) = 0$ is, of course, impossible. (This follows also from (3), (5) and Theorem 20.) The case $O(\mathfrak{W}) = 1$ clearly holds by (1) if and only if $f(\xi)$ is linear. (This could also, of course, have been obtained from Theorem 21.) Thus we have

THEOREM 24. *For a non-linear function* $f(\xi) \in \mathsf{F}$ $(\xi \in \mathsf{F})$ *the number* $O(\mathfrak{W})$ *of (different) values of the difference quotient*

$$\frac{f(\xi) - f(\eta)}{\xi - \eta} \qquad (\xi, \eta \in \mathsf{F};\ \xi \neq \eta) \tag{6}$$

*always lies in one of the intervals*

$$\left(1+\frac{q-1}{p^e+1},\ \frac{q-1}{p^e-1}\right)\left(e=1,\dots,\left[\frac{n}{2}\right]\right);\quad \left(\frac{q+1}{2},\ q\right). \tag{7}$$

This theorem is very interesting because, in spite of its primitive nature, it appears to be inaccessible by a direct approach.

When $q=p$ $(n=1)$ (7) reduces to the unique interval $\left(\frac{p+1}{2}, p\right)$. This means that the difference quotient (6) then takes at least $\frac{p+1}{2}$ values. This bound can in fact be increased by one (for $p\neq 2$); my attention was drawn to this by L. Megyesi. Indeed, the following theorem holds:

THEOREM 24′. *When $p\neq 2$ the difference quotient*

$$\frac{f(\xi)-f(\eta)}{\xi-\eta} \qquad (\xi, \eta\in \mathsf{F}_p;\ \xi\neq\eta) \tag{8}$$

*of a non-linear function $f(\xi)\in\mathsf{F}_p$ $(\xi\in\mathsf{F}_p)$ takes at least $\frac{p+3}{2}$ values.*

To prove this, we write the number of the permutations among the mappings

$$\eta \to \eta\xi - f(\eta)$$

as $N$. By the last remark and by (5) it will be sufficient to prove that $N\neq\frac{p-1}{2}$. With this in mind we assume that $N=\frac{p-1}{2}$ and form the system of linear polynomials

$$l_\eta(x) = \eta x - f(\eta).$$

As a consequence of the assumption, case 1) of Theorem 22 must arise here. As this is, however, obviously false, Theorem 24′ is proved.

We conjecture that $O(\mathfrak{W})\neq\frac{q+1}{2}$ also for $q>p$. (The proof could be expected to come from Theorem 17.)

For some simple functions $f(\xi)$ (the set $\mathfrak{W}$ and) the value of $O(\mathfrak{W})$ can be determined exactly, as is done in the following examples (except Example 3). In some of these examples we give $f(\xi)$, for convenience, in terms of a generating polynomial $f(x)\in\mathsf{F}[x]$ (of degree $\leqq q-1$). (Without loss of generality we may here confine ourselves to monic $f(x)$.) The difference quotients can be written, as usual, in the form

$$\frac{1}{h}(f(x+h)-f(x)),$$

where $x$ and $h$ are to be treated as variables in $\mathsf{F}$ and $\mathsf{F}/0$, respectively.

EXAMPLE 1. In the quadratic case

$$f(x) = x^2+\alpha x+\beta$$

the difference quotient is equal to

$$2x+h+\alpha,$$

we therefore now have

$$O(\mathfrak{W})=q.$$

EXAMPLE 2. Let $f(x)=x^{p^e}$ with $1\leqq e\leqq n-1$. The difference quotient is now

$$h^{p^e-1}$$

(it is therefore independent (!) of $x$). So clearly,

$$O(\mathfrak{W}) = \frac{q-1}{p^{(e,n)}-1}.$$

EXAMPLE 3. For

$$f(x) = \sum_{i=1}^{k} \alpha_i x^{p^{e_i}} \qquad (k \geqq 2;\ 0 \leqq e_1 < \cdots < e_k \leqq n-1;\ \alpha_1, \ldots, \alpha_k \in \mathsf{F}\setminus 0)$$

the difference quotient is equal to

$$\sum_{i=1}^{k} \alpha_i h^{p^{e_i}-1}\left(=\frac{f(h)}{h}\right).$$

Clearly

$$O(\mathfrak{W}) \leqq \frac{q-1}{p^{(e_1, \ldots, e_k, n)}-1}.$$

Obviously, for any choice of $e_1, \ldots, e_k$ there exist suitable $\alpha_1, \ldots, \alpha_k$ such that here the $<$ sign holds, but the exact determination of $O(\mathfrak{W})$ seems to be impossible in general. (Even the case $k=2$ is already very difficult.)

EXAMPLE 4. In the special case

$$f(x) = x+x^P+x^{P^2}+\cdots+x^{\frac{q}{P}} \qquad (P \underset{1}{|} q;\ P<q)$$

of Example 3 we have

$$f(\xi)=S_{\mathsf{F}\setminus \mathsf{F}_P}(\xi) \qquad (\xi\in \mathsf{F}),$$

where the right-hand side contains the symbol for the relative trace. The difference quotient

$$\frac{1}{h} S_{\mathsf{F} \setminus \mathsf{F}_P}(h) \qquad (h \in \mathsf{F} \setminus 0)$$

is invariant under the substitutions $h \to \alpha h$ $(\alpha \in \mathsf{F}_p \setminus 0)$, whence it follows very easily that we may restrict ourselves to values of $h$ with relative trace 0 or 1. Since the number of the latter $h$ is equal to $\frac{q}{P}$, it follows that

$$O(\mathfrak{W}) = 1 + \frac{q}{P}.$$

EXAMPLE 5. Let $M$ be a proper submodule of $\mathsf{F}$ and let us put

$$f(\xi) = \begin{cases} 1 & \text{for} \quad \xi \in M, \\ 0 & \text{for} \quad \xi \in \mathsf{F} \setminus M. \end{cases}$$

It is obvious that all the values of the difference quotient are now 0 and the

$$\frac{1}{\varkappa} (\varkappa \in \mathsf{F} \setminus M).$$

so that

$$O(\mathfrak{W}) = 1 + q - O(M). \tag{8}$$

(Here $O(M)$ can be any proper divisor of $q$.)

The following interesting example is due in another form to L. Megyesi:

EXAMPLE 6. Let $G$ be a proper subgroup of $\mathsf{F} \setminus 0$ and let us put

$$f(\xi) = \begin{cases} \xi & \text{for} \quad \xi \in G, \\ 0 & \text{for} \quad \xi \in \mathsf{F} \setminus G. \end{cases}$$

For $\xi, \eta \in \mathsf{F} (\xi \neq \eta)$ we have

$$\frac{f(\xi) - f(\eta)}{\xi - \eta} = \begin{cases} 1 & \text{if} \quad \xi, \eta \in G, \\ 0 & \text{if} \quad \xi, \eta \in \mathsf{F} \setminus G, \\ \dfrac{\xi}{\xi - \eta} \left( = \dfrac{1}{1 - \xi^{-1}\eta} \right) & \text{if} \quad \xi \in G, \ \eta \in \mathsf{F} \setminus G. \end{cases}$$

In the third case $\xi^{-1}\eta$ runs through the elements of $\mathsf{F} \setminus (0 \cup G)$, so that

$$O(\mathfrak{W}) = 1 + q - O(G) \tag{9}$$

holds. (Here $O(G)$ may be any proper divisor of $q-1$.) For $p \neq 2$ we can choose

$O(G) = \frac{q-1}{2}$ and therefore $\frac{q+3}{2}$ is a possible value of $O(\mathfrak{W})$. (Cf. Theorems 24, 24′.)

In this connection the following point should be noted. Let

$$x^d - \alpha (\in \mathsf{F}[x]) \qquad (d|q-1;\ 1 < d < q-1;\ \alpha \in \mathsf{F}^d \setminus 0)$$

be an Euler binomial of $d$-th degree and let us form the cofactor

$$\Phi(x) = \frac{x^q - x}{x^d - \alpha} (\in \mathsf{F}[x]).$$

(Thus $\Phi(x)$ is a lacunary fully reducible polynomial from $x\mathsf{F}[x^d]$ without multiple zeros.) The reader will see without difficulty that the function $\Phi(\xi)$ does not differ essentially from the function $f(\xi)$ considered here. (Indeed, $\varrho\Phi(\sigma\xi)$ agrees with $f(\xi)$ for suitable constants $\varrho, \sigma \in \mathsf{F} \setminus 0$.) Megyesi has obtained result (9) by the rather more complicated investigation of $\Phi(x)$.

NOTE 3. The formal resemblance of (8) and (9) is striking. We also observe that the function $f(\xi)$ given in Example 5 can be generated by a polynomial from $\mathsf{F}[x^{p-1}]$; this is easily verified by interpolation. Thus Examples 2–6 deal with functions that are all induced by (strongly) lacunary polynomials.

We were led to the following generalization by a suggestion of F. Kasch. Let $\mathfrak{W}(=\mathfrak{W}(f, g))$ be the range of the *difference quotient*

$$\frac{f(\xi)-f(\eta)}{g(\xi)-g(\eta)} \qquad (\xi, \eta \in \mathsf{F};\ g(\xi) \neq g(\eta)) \tag{10}$$

of two functions $f(\xi), g(\xi) \in \mathsf{F}$. We assume that neither $\xi \to f(\xi)$ nor $\xi \to g(\xi)$ is a permutation of $\mathsf{F}$, as otherwise after the introduction of a new variable a reduction to the above case would be possible. In order to avoid trivial cases, we further assume that neither $f(\xi)$ nor $g(\xi)$ is constant. However, we can successfully investigate $O(\mathfrak{W})$ only if we further assume that

$$[\xi, \eta \in \mathsf{F} \text{ and } \xi \neq \eta] \Rightarrow [f(\xi) \neq f(\eta) \text{ or } g(\xi) \neq g(\eta)]. \tag{11}$$

Since $g(\xi)$ is not constant, we again have $O(\mathfrak{W}) \geqq 1$, but first we shall show that

$$O(\mathfrak{W}) > 1. \tag{12}$$

We assume that $O(\mathfrak{W}) = 1$ and write the unique value of (10) as $\alpha$. Then

$$[\xi, \eta \in \mathsf{F} \quad \text{and} \quad g(\xi) \neq g(\eta)] \Rightarrow f(\xi) - \alpha g(\xi) = f(\eta) - \alpha g(\eta). \tag{13}$$

We shall show that the equation

$$f(\xi) - \alpha g(\xi) = f(\eta) - \alpha g(\eta) \tag{14}$$

holds for all $\xi, \eta \in \mathsf{F}$. From (13) it follows that we need now only consider $g(\xi)=g(\eta)$. From the assumption there exists a $\zeta \in \mathsf{F}$ with $g(\xi)(=g(\eta)) \neq g(\zeta)$. The application of (13) to both pairs $\xi, \zeta$ and $\eta, \zeta$ (in place of $\xi, \eta$) gives that both sides of (14) are equal to $f(\zeta)-\alpha g(\zeta)$, and therefore also to one another. This proves (14) for all $\xi, \eta \in \mathsf{F}$. Now the assumption implies the existence of elements $\xi, \eta \in \mathsf{F}$ such that $\xi \neq \eta$ and $g(\xi)=g(\eta)$. Since $f(\xi)=f(\eta)$ follows from this and from (14), we have obtained a contradiction to (11), which proves (12).

We further prove the following analogue of Proposition 46: A $\zeta \in \mathsf{F}$ belongs to the range $\mathfrak{W}$ of the difference quotient (10) if and only if

$$\eta \to g(\eta)\zeta - f(\eta) \qquad (\eta \in \mathsf{F}) \tag{15}$$

is not a permutation.

It is necessary and sufficient for $\zeta \notin \mathfrak{W}$ that

$$g(\xi)\zeta - f(\xi) \neq g(\eta)\zeta - f(\eta)$$

should follow from $\xi, \eta \in \mathsf{F}$ and $g(\xi) \neq g(\eta)$. However, this inequality also holds by (11) when $g(\xi)=g(\eta)$ with $\xi \neq \eta$ and therefore, together with (11), it is equivalent to the statement that (15) is a permutation. The assertion follows from this by a logical inversion.

Proceeding as above we obtain

THEOREM 24″. *If $f(\xi)$ and $g(\xi)$ are non-constant functions in* F *such that neither $\xi \to f(\xi)$ nor $\xi \to g(\xi)$ is a permutation of* F *and the inclusion* (11) *holds then the cardinal number $O(\mathfrak{W})$ of the range $\mathfrak{W}=\mathfrak{W}(f, g)$ of the difference quotient* (10) *lies in one of the intervals indicated at* (7).

It is difficult to construct examples for this theorem in which $O(\mathfrak{W})<q$. The following is such an example

EXAMPLE 7. Let $\mathsf{F} = \mathsf{A} \oplus \mathsf{B}$ be the decomposition of the module of F into the direct sum of two proper submodules A and B. Then

$$O(\mathsf{A})\,O(\mathsf{B})=q.$$

We define $f(\xi)$ and $g(\xi)$ in such a way that

$$\xi=f(\xi)+g(\xi) \qquad (\xi \in \mathsf{F},\ f(\xi) \in \mathsf{A},\ g(\xi) \in \mathsf{B}).$$

(In other words: $\xi \to f(\xi)$ and $\xi \to g(\xi)$ are the projections of F in A and in B, respectively.) The premises of Theorem 24″ are clearly fulfilled for those functions. It is also clear that $\mathfrak{W}$ consists of all quotients

$$\frac{\alpha}{\beta} \qquad (\alpha \in \mathsf{A};\ \beta \in \mathsf{B} \setminus 0).$$

Since a common factor 1, ..., $p-1$ can be inserted in the numerator and the denominator, the estimation

$$O(\mathfrak{W}) \leqq 1+\frac{(O(\mathsf{A})-1)(O(\mathsf{B})-1)}{p-1}$$

follows.

According to a private communication from E. Fried the stronger estimation

$$O(\mathfrak{W}) \leqq 1+\frac{(O(\mathsf{A})-1)(O(\mathsf{B})-1)}{O(\mathsf{L})-1}$$

trivially holds if $\mathsf{L}$ is a proper subfield of $\mathsf{F}$ and $\mathsf{A}$ and $\mathsf{B}$ are restricted to being $\mathsf{L}$-modules.

## § 37. Common representative systems of a finite abelian group with respect to given subgroups

Let $G$ be a finite abelian group. The complexes of $G$ containing the unity element are said to be *normed*. The transition from a complex $C$ to the complex $\varkappa^{-1}C$ with arbitrary $\varkappa \in C$ is called a *norming* of $C$. (The latter has already been applied in Proposition 19.)

We assume that different subgroups $H_1, \ldots, H_k$ ($k \geqq 1$) of $G$ are given and ask whether $G$ then possesses a *common normed representative system* with respect to all the $H_1, \ldots, H_k$ which is not a group. For a common representative system to exist at all, $O(H_1) = \cdots = O(H_k)$ must, of course, hold. It follows from this for a cyclic $G$ that $k=1$; thus the question raised is interesting only for noncyclic groups $G$, but it is then very difficult.

We have been able to investigate only the case of an elementary $G$ such that $O(G)=p^2$. The case $p=2$ is devoid of interest; thus we assume $p \neq 2$. We also assume $O(H_i)=p$ ($i=1, \ldots, k$), as otherwise we are dealing with a trivial case. A common representative system $R$ of $G$ with respect to all the $H_1, \ldots, H_k$ is characterized by the validity of all schlicht product decompositions

$$G = R \cdot H_i \qquad (i = 1, \ldots, k). \tag{1}$$

We assume (1) and $1 \in R$ and that $R$ is not a group, in order to deduce necessary conditions from these assumptions. If $k = p+1$, i.e. the $H_i$ are all the subgroups of order $p$, then their union is equal to $G$, so that (1) is then impossible. It is therefore sufficient to consider the case

$$k \leqq p. \tag{2}$$

Let us denote by $\langle \gamma \rangle$ the subgroup generated by a $\gamma \in G$. It follows from (2) that it is possible to choose a basis $\alpha$, $\beta$ of $G$ such that ($\langle\beta\rangle$ does not appear among the $H_1, \ldots, H_k$, i.e.) we may write

$$H_i = \langle \alpha\beta^{-h_i} \rangle \qquad (i=1, \ldots, k). \tag{3}$$

(The numbers $h_1, \ldots, h_k$ are different mod $p$, of course.) As $O(R)=p$ must hold, we may put

$$R = \{\alpha^{a_0}\beta^{b_0}, \ldots, \alpha^{a_{p-1}}\beta^{b_{p-1}}\} \qquad (a_0 = b_0 = 1). \tag{4}$$

Let $\varrho \neq 1$ be a complex $p$-th root of unity. We shall use the $p$ characters $\omega_0, \ldots, \omega_{p-1}$ of $G$ defined by

$$\omega_c(\alpha^a \beta^b) = \varrho^{ac+b} \qquad (c = 0, \ldots, p-1). \tag{5}$$

It follows then from (1), (3), (4), (5) that

$$0 = \sum_{j=0}^{p-1} \varrho^{a_j c + b_j} \sum_{j=0}^{p-1} \varrho^{j(c-h_i)}$$

holds for all $i=1, \ldots, k$ and $c=0, \ldots, p-1$. We apply this equation whith $c \equiv h_i \pmod p$; then the second factor of the right-hand side does not vanish (it is actually equal to $p$), so that

$$\sum_{j=0}^{p-1} \varrho^{a_j h_i + b_j} = 0 \qquad (i=1, \ldots, k)$$

follows. Putting

$$l_j(x) = a_j x + b_j \qquad (j = 0, \ldots, p-1) \tag{6}$$

and interpreting this polynomial in $\mathsf{F}_p[x]$, it follows that the $k$ mappings

$$j \to l_j(h_i) = a_j h_i + b_j \qquad (i = 1, \ldots, k) \tag{7}$$

of $\mathsf{F}_p$ into itself are permutations.

We consider all the $p$ mappings

$$j \to l_j(h) = a_j h + b_j \qquad (h = 0, \ldots, p-1) \tag{8}$$

of $\mathsf{F}_p$ into itself and denote by $N$ the number of the permutations appearing among them. From (7) we then have

$$N \geqq k; \tag{9}$$

on the other hand, $N$ must fall into one of the intervals

$$\left[0, \frac{p-1}{2}\right], \quad (p-1, p), \tag{10}$$

according to the case $q=p$ of Theorem 19. We show that the cases $N=p$ or $p-1$ are impossible.

For, if $N=p$, then by Theorem 20 (for $q=p$) it follows for the coefficients appearing in (6) that the $a_0, \ldots, a_{p-1}$ are congruent mod $p$ to one another and the $b_0, \ldots, b_{p-1}$ constitute a representative system mod $p$ (of $\mathscr{Z}$). This means by (4) that $R$ is a subgroup of $G$.

If $N = p-1$, then it follows similarly from (6) and from (the case $q=p$ of) Theorem 21 that the $a_0, \ldots, a_{p-1}$ form a representative system mod $p$ and satisfy the congruences

$$b_j \equiv -ra_j+s \qquad (\text{mod}\, p) \quad (j = 0, \ldots, p-1)$$

for suitable $r, s \in \mathscr{Z}$. Because $a_0=b_0=0$ this holds even for $s=0$. This means by (4) that $R$ consists of the elements

$$\alpha^{a_j}\beta^{-ra_j}(= (\alpha\beta^{-r})^{a_j}) \qquad (j = 0, \ldots, p-1),$$

so that it is again a subgroup of $G$. Since, however, we have assumed that $R$ is not a group, this proves the assertion $N \neq p$ or $p-1$.

This implies that

$$0 \leqq N \leqq \frac{p-1}{2}. \tag{11}$$

By (9) we have thus obtained the result that a desired representative system $R$ can exist only when

$$k \leqq \frac{p-1}{2}. \tag{12}$$

Before giving the definitive formulation of this, let us consider more thoroughly the case where, in (12), the $=$ sign holds. By virtue of (9) and (11) we then have $N = \frac{p-1}{2}$. This means that Case 1) of Theorem 22 can be applied to the system (6).

Before dealing with this we note that the system (6) may be replaced by

$$l_j(x) - l_i(x) \qquad (j = 0, \ldots, p-1),$$

where $l_i(x)$ is an arbitrarily chosen fixed element from (6), as the only consequence of this replacement is that $R$ is normed in another way.

It follows that Case 1) of Theorem 22 with $a(x)=0$ can be applied to (6). This means that the polynomials (6) are, apart from their order, the following:

$$\begin{cases} 0, & \\ r(ax+b) & \left(\left(\frac{r}{p}\right) = 1\right), \\ s(cx+d) & \left(\left(\frac{s}{p}\right) = -1\right), \end{cases} \tag{13}$$

where $a$, $b$, $c$, $d$ are fixed integers (also treated as elements of $\mathsf{F}_p$) so that we have, *inter alia*, that

$$p \nmid ad - bc \tag{14}$$

and $r$, $s$, as pointed out already, run through the quadratic residues and non-residues mod $p$, respectively (from $\mathscr{L}$). Comparison with (4) shows that $R$ consists of the elements

$$1; \quad (\alpha^a \beta^b)^r \left(\left(\frac{r}{p}\right) = 1\right); \quad (\alpha^c \beta^d)^s \left(\left(\frac{s}{p}\right) = -1\right).$$

By (14) the power products in the brackets can be chosen as new basic elements of $G$ for which we shall retain the notation $\alpha$, $\beta$. Then, what we have said above, means that when $k = \dfrac{p-1}{2}$

$$R = \left\{1, \alpha^r, \beta^s: \quad \left(\frac{r}{p}\right) = 1, \quad \left(\frac{s}{p}\right) = -1\right\} \tag{15}$$

(where the notation is self-evident).

Conversely, let us assume (15) and determine all subgroups $H$ of $G$ such that

$$G = R \cdot H. \tag{16}$$

Of course we may here assume that

$$O(H) = p.$$

For $H = \langle \alpha \rangle$ or $H = \langle \beta \rangle$ (16) is obviously false, so that $H$ must be of the form

$$H = \langle \alpha \beta^{-t} \rangle, \tag{17}$$

where $p \nmid t$. If $\left(\dfrac{t}{p}\right) = -1$, then for the elements

$$\varrho_1 = \beta^t, \quad \varrho_2 = \alpha$$

of $R$ and for the elements

$$\varkappa_1 = \alpha \beta^{-t}, \quad \varkappa_2 = 1$$

of $H$ we have the equation $\varrho_1 \varkappa_1 = \varrho_2 \varkappa_2$. Thus (16) is again false. Therefore

$$\left(\frac{t}{p}\right) = 1 \tag{18}$$

must hold.

Conditions (17) and (18) are sufficient for the validity of (16). This follows from the fact that, from (15), (17) and (18), there is no element

$$\varrho_1 \varrho_2^{-1} \qquad (\varrho_1, \varrho_2 \in R;\ \varrho_1 \neq \varrho_2)$$

in $H$.

We have proved that in case (15) equation (16) is satisfied exactly only for those subgroups $H$ that are determined by (17) and (18). It follows moreover that the number of these $H$ is $\frac{p-1}{2}$.

Thus we have proved

THEOREM 25. *In an elementary abelian group G of order* $p^2$ $(p \neq 2)$ $\frac{p-1}{2}$ *is the maximum number of those subgroups (of order p) for which G possesses with respect to these subgroups a common representative system which is normed and not a group. For* $\frac{p-1}{2}$ *subgroups* $H_1, \ldots, H_{\frac{p-1}{2}}$ *the group G has such a common representative system R if and only if, for a suitable choice of a basis* $\alpha$, $\beta$ *of G, these subgroups are (to within a rearrangement) the cyclic groups generated by the single elements* $\alpha\beta^{-i^2}$ $\left(i=1, \ldots, \frac{p-1}{2}\right)$. *Moreover, R is then uniquely determined (apart from other normings) and can be chosen as the set composed of the elements*

$$1;\quad \alpha^r\left(\left(\frac{r}{p}\right)=1\right);\quad \beta^s\left(\left(\frac{s}{p}\right)=-1\right).$$

## § 38. Divisibility maximum properties of the gaussian sums and related questions

We denote the *p-th cyclotomic field* over the rational number field by $\mathfrak{R}_p$. This field can be generated by any complex $p$-th root of unity $\neq 1$. We choose a fixed one of these and denote it by $\varrho$. We write the ring of the integer elements of $\mathfrak{R}_p$ as $[\mathfrak{R}_p]$. As is well-known, the elements

$$1, \varrho, \ldots, \varrho^{p-2}$$

form an *integral basis* of $\mathfrak{R}_p$, i.e. a basis of $[\mathfrak{R}_p]$. This means that

$$\alpha = a_0 + a_1\varrho + \cdots + a_{p-2}\varrho^{p-2} \qquad (a_0, \ldots, a_{p-2} \in \mathscr{Z}) \tag{1}$$

is the (unique) basis representation of the elements of $[\mathfrak{R}_p]$. It should be noted that as

$$1 + \varrho + \cdots + \varrho^{p-1} = 0$$

(1) may be easily replaced by

$$\alpha = a_0 + a_1\varrho + \cdots + a_{p-1}\varrho^{p-1} \quad (a_0, \ldots, a_{p-1} \in \mathcal{N}), \tag{2}$$

this representation being likewise unique if the coefficients satisfy the condition

$$\min(a_0, \ldots, a_{p-1}) = 0.$$

The transition from (2) to (1) is even easier.

For an element $\alpha$ of the ring $[\mathfrak{R}_p]$, by the "ideal $\alpha$" we understand, as usual, the principal ideal of this ring generated by $\alpha$. It is well known that the ideal $1-\varrho$ is the only prime divisor of $p$ and that

$$(1-\varrho)^{p-1} \| p. \tag{3}$$

Since, further, the discriminant of $[\mathfrak{R}_p]$ is equal to $\left(\frac{-1}{p}\right) p^{p-2}$, $1-\varrho$ is the only ramified prime ideal of $[\mathfrak{R}_p]$.

In view of the importance of the prime ideal $1-\varrho$ it is interesting to calculate the *exponent* $e = e(\alpha)$ of $1-\varrho$ in $\alpha$ for an $\alpha \in [\mathfrak{R}_p] \setminus 0$, where

$$(1-\varrho)^e \| \alpha \tag{4}$$

is taken as the definition of $e$. We shall now deal with this.

If $N(\alpha)$ denotes the norm of $\alpha$ in $\mathfrak{R}_p$, then, of course,

$$p^e \| N(\alpha)$$

follows from (3) and (4). This relation can be considered as a solution of our problem, nevertheless this method is in general bound up with serious difficulties in counting. We shall therefore look for another way.

Our procedure consists first of all of reducing the problem to the case $e \leqq p-2$ and then solving it with the aid of representation (2).

By virtue of (1), $p^k | \alpha$ holds if and only if

$$p^k | a_0, \ldots, a_{p-2}.$$

Then (1) implies the basis representation

$$\frac{\alpha}{p^k} = \frac{a_0}{p^k} + \cdots + \frac{a_{p-2}}{p^k}\varrho^{p-2}$$

and

$$e = e\left(\frac{\alpha}{p^k}\right) + k(p-1)$$

holds by virtue of (3). Application to the case $p^k \| \alpha$ gives the required reduction.

For the reduced case the solution of our problem is given by

LEMMA III. *Let* $\varrho$ *be a complex p-th root of unity* $\neq 1$ *and*

$$\alpha = a_0 + a_1\varrho + \cdots + a_{p-1}\varrho^{p-1} \quad (a_0, \ldots, a_{p-1} \in \mathcal{N}) \tag{5}$$

*be an (arbitrary) integer element of the p-th cyclotomic field* $\mathfrak{R}_p$. *Then the divisibility*

$$1 - \varrho | \alpha \tag{6}$$

*is equivalent to*

$$p | a_0 + \cdots + a_{p-1}. \tag{7}$$

*In this case, for a natural number* $t \leqq p-1$, *the divisibility*

$$(1-\varrho)^t | \alpha \tag{8}$$

*is equivalent to the fact that* (7) *holds and for the sinking of the polynomial*

$$f(x) = \prod_{i=0}^{p-1} (x-i)^{a_i} \in \mathsf{F}_p[x] \tag{9}$$

*the lower estimate*

$$f^{\downarrow} \geqq t \tag{10}$$

*holds.*

COROLLARY. *For a natural number* $t \leqq p-2$

$$(1-\varrho)^t \| \alpha \tag{11}$$

*holds if and only if* (7) *and*

$$f^{\downarrow} = t \tag{12}$$

*hold.* (*For* $t=0$ (11) *holds if and only if* (7) *is false.*)

The first part of Lemma III is clearly true, as $\varrho^i \equiv 1 \pmod{1-\varrho}$. We shall prove the second part by making appropriate changes in its formulation, writing (5) in the form

$$\alpha = \sum_{j=1}^{r} \varrho^{k_j} \qquad (r = a_0 + \cdots + a_{p-1}), \tag{13}$$

where each $k_j$ is one of the elements $0, \ldots, p-1$ and every $i=0, \ldots, p-1$ appears exactly $a_i$ times among the $k_1, \ldots, k_r$. Since (6) is now assumed to hold (7), i.e.

$$p | r \tag{14}$$

also holds as a consequence of the part of the lemma that we have already proved. It should be noted that [after the transition from (5) to (13)] (9) assumes the form

$$f(x) = (x-k_1) \ldots (x-k_r), \tag{15}$$

using (13). If therefore $s_l(k_1, \ldots, k_r)$ denotes the $l$-th elementary symmetrical expression of $k_1, \ldots, k_r$, then condition (10) is equivalent to

$$s_l(k_1, \ldots, k_r) = 0 \qquad (l = 1, \ldots, t-1).$$

When $t \leqq p-1$ this system of equations is, by Newton's formulae, equivalent to the system

$$k_1^l + \cdots + k_r^l = 0 \qquad (l = 1, \ldots, t-1)$$

and therefore clearly also to the system

$$\binom{k_1}{l} + \cdots + \binom{k_r}{l} = 0 \qquad (l = 1, \ldots, t-1). \tag{16}$$

Finally, it should be noted that (14) may be written as

$$\binom{k_1}{0} + \cdots + \binom{k_r}{0} = 0. \tag{17}$$

(This equation, as well as the system (16), is to be interpreted in $\mathsf{F}_p$.) Thus (16) and (17) are summarized in the system

$$\sum_{j=1}^{r} \binom{k_j}{l} = 0 \qquad (l = 0, \ldots, t-1). \tag{18}$$

Hence the second part of Lemma III can be reformulated as follows:

For arbitrary $k_1, \ldots, k_r$ $(=0, \ldots, p-1)$ and a natural number $t \leqq p-1$, the divisibility (8), which bears on (13), is equivalent to the system of equations (18) (in $\mathsf{F}_p$).

The proof of this assertion will be easy. By the above the assertion is correct for $t=1$. If $t \geqq 2$ we assume its validity for $t-1$. Let us break the condition (8) up into the two partial conditions

$$1-\varrho \mid \alpha, \quad (1-\varrho)^{t-1} \,\Big|\, \frac{\alpha}{1-\varrho}. \tag{19}$$

The first part of these conditions, $(19_1)$, is equivalent to $p \mid r$, by the part of the lemma already proved. Thus $(19_2)$ can be replaced, using $1 \leqq t \leqq p-1$, by

$$(1-\varrho)^{t-1} \,\Big|\, \frac{\alpha - r}{1-\varrho}. \tag{20}$$

By virtue of (13) the right-hand side takes the form

$$\left(\frac{1}{1-\varrho} \sum_{j=1}^{r} \varrho^{k_j} - 1) = \right) - \sum_{j=1}^{r} \sum_{i=0}^{k_j - 1} \varrho^i,$$

so that, by the induction hypothesis, (20) is equivalent to

$$\sum_{j=1}^{r}\sum_{i=0}^{k_j-1}{}'(k_{j_i}-1)=0 \qquad (l=0,\ldots,t-2),$$

and this can be replaced by

$$\sum_{j=1}^{r}\binom{k_j}{l+1}=0 \qquad (l=0,\ldots,1-2),$$

i.e.

$$\sum_{j=1}^{r}\binom{k_j}{l}=0 \qquad (l=1,\ldots,t-1). \tag{21}$$

We have thus shown that $(19_2)$ may be replaced by (21). On the other hand, as was pointed out above, $(19_1)$ is equivalent to

$$\sum_{j=0}^{r}\binom{k_j}{0}=0,$$

so that (19) is equivalent to (18). This completes the proof of Lemma III. The corollary follows immediately, from the lemma. (For another proof of this lemma see Carlitz [1], p. 140.)

We see from the corollary that our problem can be completely reduced to fully reducible lacunary polynomials in $\mathsf{F}_p[x]$ (the degree of which is a multiple of $p$, except for a few simple cases).

For our first application, let us consider for $p\neq 2$ the $p$-termed sums

$$S=\varrho^{k_1}+\cdots+\varrho^{k_p} \tag{22}$$

of complex $p$-th roots of unity. We disregard the case $S=0$. This means that at least two equal terms appear in $S$. Let

$$(1-\varrho)^e\|S. \tag{23}$$

We shall try to find out something about the possible values of this number $e$.

In the case $e\geqq p-1$ it follows from (3) and (23) that $p|S$, whence on the strength of (22) and $S\neq 0$ we infer that all the terms of $S$ are equal; thus $S$ is associated with $p$ and $e=p-1$. We have thus proved that

$$e\leqq p-1$$

must hold. As, however, the case $e=p-1$ is trivial, we have only the case

$$e\leqq p-2 \tag{24}$$

left to solve. We can now use the corollary of Lemma III. For this purpose we identify (22) with the special case $a_0+\cdots+a_{p-1}=p$ of (5). The polynomial (9) (or (15)) becomes

$$f(x) = (x-k_1)\ldots(x-k_p). \tag{25}$$

Condition (7) is satisfied now, so, by the corollary we have

$$f^{\downarrow}=e. \tag{26}$$

If, now,

$$e \geqq \frac{p-1}{2}, \tag{27}$$

then $f^{\circ\circ} \leqq \dfrac{p+1}{2}$ follows from (26) and $f^{\circ}=p$; therefore $f(x)$ is a solution of the case $q=p$ of Problem II. So, by Theorem 17, either

$$f(x) = x^p - x$$

or (cf. Example 1, which comes after Theorem 17)

$$f(x) = \left((x+a)(x+a)^{\frac{p-1}{2}}-\sigma\right)\left((x+a)^{\frac{p-1}{2}}-\sigma\tau\right) \tag{28}$$

$$(a\in \mathsf{F}_p;\ \sigma=\pm 1;\ \tau=0, 1).$$

(In reality, $a$ may be any integer which, however, is to be interpreted in (28) as an element of $\mathsf{F}_p$.)

When $f(x) = x^p-x$ we have $S=0$, by (22) and (25). Since, however, we have assumed that $S\neq 0$, (28) must hold. This may be written as

$$f(x-a) = x\left(x^{\frac{p-1}{2}}-\sigma\right)\left(x^{\frac{p-1}{2}}-\sigma\tau\right).$$

If we replace $S$ by $\varrho^a S$, then $f(x)$ clearly becomes $f(x-a)$, so that we may assume without any substantial loss of generality that

$$f(x) = x\left(x^{\frac{p-1}{2}}-\sigma\right)\left(x^{\frac{p-1}{2}}-\sigma\tau\right) \qquad (\sigma=\pm 1;\ \tau=0, 1). \tag{29}$$

It should be noted that for (29) $f^{\circ\circ} = \dfrac{p+1}{2}$, i.e. $f^{\downarrow} = \dfrac{p-1}{2}$ always holds. From this and from (26) it follows that we have thus shown that in (27) the $=$ sign must hold.

Now we can see by (29) that the following four cases are possible:

$$f(x) = x\left(x^{\frac{p-1}{2}}-1\right)^2,\quad x\left(x^{\frac{p-1}{2}}+1\right)^2,\quad x^{\frac{p+1}{2}}\left(x^{\frac{p-1}{2}}-1\right),\quad x^{\frac{p+1}{2}}\left(x^{\frac{p-1}{2}}+1\right).$$

The corresponding four values of $S$ are denoted by $S_1, \ldots, S_4$. Comparison with (15) and (22) shows easily that these can be expressed by the gaussian sum

$$\Gamma = \sum_{i=0}^{p-1} \varrho^{i^2} \left( = \sum_{i=0}^{p-1} \left(\frac{i}{p}\right) \varrho^i \right)$$

in the following way:

$$S_1 = \Gamma, \quad S_2 = -\Gamma, \quad S_3 = \frac{1}{2}(p+\Gamma), \quad S_4 = \frac{1}{2}(p-\Gamma).$$

We have thus proved

THEOREM 26. *If $S \neq 0$ is a p-termed sum of complex p-th roots of unity, not all equal (so that $p \neq 2$), and $\varrho \neq 1$ is a complex p-th root of unity, then S is divisible at most by the $\frac{p-1}{2}$-th power of $1-\varrho$ and it is divisible by this power if and only if S has one of the four values*

$$\pm\Gamma, \quad \frac{1}{2}(p \pm \Gamma), \tag{30}$$

*apart from a norming factor $\varrho^a$.*

This state of affairs is called the *divisibility maximum property* of the gaussian sum $\Gamma$ (within the "non-trivial" $p$-termed sums of $p$-th complex roots of unity).

NOTE 1. The "if" part of Theorem 26 also follows from the known equation

$$\Gamma^2 = \left(\frac{-1}{p}\right) p.$$

As a further application, we consider non-vanishing sums of the form

$$S = \pm\varrho^{l_1} \pm \cdots \pm \varrho^{l_r} \qquad (1 \leqq r \leqq p-1) \tag{31}$$

with $p \neq 2$, where we allow arbitrary signs. These $S$ are clearly not divisible by the $p-1$-th power of $1-\varrho$. It is easy to see that (for $r = p-1$) (31) takes, among others, the four values (30). The question therefore arises whether $\Gamma$ retains its divisibility maximum property even within the sums (31). (That is to say, whether Theorem 26 is valid for the sums (31).) We have been unable to answer this question in full generality. The situation is changed if in (31) the terms with + sign or those with − sign are different from each other, which will be assumed to be the case from now on. Each case turns into the other if we interchange $S$ and $-S$. We shall therefore stick to the second case, i.e. where in (31) the terms of the form $-\varrho^l$ are all different. Then we can easily see that the answer to our question is "yes". Moreover we shall show that the exponent of $1-\varrho$ in (31) is then at most equal to $\frac{r}{2}$.

Add $1+\varrho+\cdots+\varrho^{p-1}(=0)$ to the right-hand side of (31) and then write $S'$ instead of $S$. Because $S'=S$, this change is unessential, and it follows from the assumption made in the meantime that $S'$ will appear among the sums admitted in Theorem 26. This implies the first assertion.

The proof of the second assertion will come from a careful consideration of $S'$. We point out, in advance, that when $1-\varrho|S$ clearly $2|r$ must hold, by (31), and, moreover, an equal number of $+$ and $-$ signs must appear on the right-hand side of (31). It will therefore be enough to consider only this case. Then $S'$ is the sum of $p$ terms which are $p$-th roots of unity, containing at least $p-\frac{r}{2}$ distinct terms. Now let us form the polynomial (25) for $S'$ (instead of the sum given in (22)) and denote it by $g(x)$. (We have $g(x)\underset{1}{\in}\mathsf{F}_p[x]$, $g^{\circ}=p$ and $g(x)$ is fully reducible.) It follows, on the one hand, that $g(x)$ has at least $p-\frac{r}{2}$ different zeros (in $\mathsf{F}_p$) and therefore the same holds for

$$g(x)-(x^p-x).$$

The degree of these polynomials, i.e. the second degree $g^{\circ\circ}$ of $g(x)$ is thus greater than or equal to $p-\frac{r}{2}$. (An exception can occur, but only for $p-\frac{r}{2}=1$, and then $p=3$ (because $r\leqq p-1$), a case which we shall disregard for the time being.) Thus (as $g^{\circ}=p$)

$$g^{\downarrow}\leqq\frac{r}{2}.$$

On the other hand, let us put

$$(1-\varrho)^e\|S \qquad (\text{i.e. } (1-\varrho)^e\|S').$$

As, according to the assumption, $e\geqq 1$ and $p\nmid S'$, i.e. $e\leqq p-2$, it follows from the corollary of Lemma III (applied to $S'$ instead of $\alpha$) that

$$g^{\downarrow}=e.$$

From these two results we obtain $e\leqq\frac{r}{2}$, which proves the second assertion for $p\neq 3$. The case $p=3$ disregarded up to now is not an exception, as then $e=1$, $r=2$. Thus we have proved

THEOREM 27. *Let $\varrho\neq 1$ be a complex p-th root of unity ($p\neq 2$). For a non-vanishing sum*

$$S=\pm\varrho^{l_1}\pm\cdots\pm\varrho^{l_r} \qquad (1\leqq r\leqq p-1) \tag{32}$$

*let us assume that all its terms of the form* $-\varrho^{l_i}$ *are different. Then for the validity of* $1-\varrho|S$ *it is necessary and sufficient that* $2|r$ *should hold and that in* (32) *the number of + signs be equal to that of − signs. In this case S is divisible by at most the* $\frac{r}{2}$*-th power of* $1-\varrho$ *and it is divisible by the* $\frac{p-1}{2}$*-th power of* $1-\varrho$ *if and only if* ($r=p-1$ *and*) *S has one of the four values* (30).

Let us consider the special cases

$$S=\varrho\pm\varrho^2\pm\cdots\pm\varrho^{p-1}\left(\neq\Gamma,\ \frac{1}{2}(p+\Gamma)\right) \tag{33}$$

of (32) more closely (having disregarded the cases that we already know). We again put

$$(1-\varrho)^e\|S. \tag{34}$$

By virtue of Theorem 27 we now certainly have

$$e<\frac{p-1}{2}, \tag{35}$$

but we shall show further that

$$e\leqq\frac{p-1}{4}. \tag{36}$$

We shall also determine precisely the cases for which equality holds in (35).

With this in mind, we retain the notations $S'$ and $g(x)$ used in the proof of Theorem 27. Since $S'$ is now of the form

$$S'=1+2\varrho+(1\pm1)\varrho^2+\cdots+(1\pm1)\varrho^{p-1}, \tag{37}$$

by (33) we may write

$$g(x)=xh(x)^2, \tag{38}$$

where

$$h(x)\underset{1}{\in}\mathsf{F}_p[x],\quad h^{\circ}=\frac{p-1}{2},\quad h(x)|x^{p-1}-1,\quad x-1|h(x). \tag{39}$$

We infer from (34) and the corollary of Lemma III that $g^{\downarrow}=e$, so that $h^{\downarrow}=e$, i.e. that

$$h^{\circ\circ}=\frac{p-1}{2}-e. \tag{40}$$

We shall assume that

$$e\geqq\frac{p-1}{4}. \tag{41}$$

Then, from (39) and (40), $h(x)$ is a solution of Problem I for $q=p$, $d=2$; so, by Theorem 5, we need consider only

$$h(x) = x^{\frac{p-1}{2}} \pm 1 \tag{42}$$

and, if $4|p-1$,

$$h(x) = \left(x^{\frac{p-1}{4}} - \beta\right)\left(x^{\frac{p-1}{4}} - \gamma\right) \qquad (\beta^2 = 1,\ \gamma^2 = -1) \tag{43}$$

In the case (42) $S'$ is equal to $\Gamma$ or $-\Gamma$, because of (38), but (as $S=S'$) this is in contradiction to our assumption. Accordingly $4|p-1$, and (43) must hold. So we see that

$$h^{\circ\circ} = \frac{p-1}{4}.$$

Then, by (40), equality must hold in (41). This proves, in the first place, that

$$e \leqq \frac{p-1}{4}$$

always holds. It also follows from (39) that in (43) only $\beta=1$ can arise, so that

$$h(x) = \left(x^{\frac{p-1}{4}} - 1\right)\left(x^{\frac{p-1}{4}} - \gamma\right) \qquad (\gamma^2 = -1). \tag{44}$$

In order to give expressions for the sums belonging to the cases (44) which are thus the only ones with $e = \dfrac{p-1}{4}$, we consider for an arbitrary divisor

$$d|p-1$$

all the Euler binomials of degree $d$. These polynomials are:

$$f_{di}(x) = x^d - \delta^i \qquad (i = 0, \ldots, d-1), \tag{45}$$

where ($di$ is not a product and) $\delta$ is an arbitrarily chosen fixed element of order $\dfrac{p-1}{d}$ of the group $\mathsf{F}_p \setminus 0$. We put

$$\Gamma_{di} = \sum_{f_{di}(k)=0} \varrho^k \qquad (i = 0, \ldots, d-1). \tag{46}$$

(The summation is to be carried out over all the zeros $k \in \mathsf{F}_p \setminus 0$ of (45) which are to be treated—in the exponent—as elements of $\mathscr{Z}$.) These $\Gamma_{di}$ are the so-called *d-termed gaussian periods*. In particular

$$\Gamma = 1 + 2\Gamma_{\frac{p-1}{2}0}.$$

follows from the corollary of Lemma III, applied to (53). As, however, $g^{\downarrow}=f^{\downarrow}(=r-f^{\circ\circ})$ can be deduced from (56) and Proposition 4, it follows that

$$e = r - f^{\circ\circ}.$$

Accordingly, in cases 1), 2) and 3) we have

$$e < r - \frac{r^2}{p-1}, \quad e = r \quad \text{and} \quad e = \frac{p-1}{4}\left(=\frac{r}{2} = r - \frac{r^2}{p-1}\right),$$

respectively.

In connection with Case 2), we take into account definition (46); then we may formulate (with $d$ in place of $r$) a part (not contained in Theorems 27 and 28) of the results obtained as follows:

THEOREM 29. *Of the sums* (32) *in Theorem 27, consider only the special cases*

$$S = \sum_{i=1}^{d}(1-\varrho^{k_i})\left(= d - \sum_{i=1}^{d}\varrho^{k_i}\right)$$

$$(d|p-1;\, d < p-1;\; 1 \leqq k_1 < \cdots < k_d \leqq p-1).$$

*All the*

$$d-\Gamma_{di} \qquad (i = 0, \ldots, d-1)$$

*appear among these, where the $\Gamma_{di}$ are the d-termed gaussian periods, and, for them,*

$$(1-\varrho)^d \| d-\Gamma_{di}$$

*holds. For the exponent of $1-\varrho$ in the remaining $S$ the inequality*

$$e \leqq d - \frac{d^2}{p-1}$$

*holds.*

It should be noted that this theorem also refers to a divisibility maximum property involving all the gaussian periods.

NOTE 3. L. Carlitz [1] has proved the above results in another way for the relatively simple sums (cf. (33))

$$S = \varrho \pm \varrho^2 \pm \cdots \pm \varrho^{p-1}$$

(without using Theorems 5, 17) where certain lacunary polynomials from $\mathsf{F}_p[x]$ have also been used.

NOTE 4. In the same paper L. Carlitz proved a number of interesting results of a very different kind, which come partly from applications of the above results and partly from an investigation of other related questions.

## § 39. Homogeneous elementary symmetric systems of equations in finite fields

Although the applications to be discussed here are less important, they are not wholly devoid of interest. They will derive from part of Theorems 17 and 18 by a simple reformulation.

We shall use the abbreviation $X_l$ ($l=1, 2, \ldots$) for a sequence $x_1, \ldots, x_l$. If all the terms of $X_l$ lie in $\mathsf{F}$ then we shall write $X_l \in (\mathsf{F}, \mathsf{F}, \ldots)$. On the other hand, we understand by $\mathsf{F}[X_l]$ the polynomial ring $\mathsf{F}[x_1, \ldots, x_l]$ (with the independent generating indeterminates $x_1, \ldots, x_l$ over $\mathsf{F}$). Any system of (algebraic) equations with $l$ un-knowns in $\mathsf{F}$ can be written in the form

$$f_i(X_l)=0 \qquad (f_i(X_l) \in \mathsf{F}[X_l];\ i=1, \ldots, k). \tag{1}$$

The $X \in (\mathsf{F}, \mathsf{F}, \ldots)$, satisfying this system, are called its *solutions*. (1) is called *symmetric* if the left-hand sides are symmetric polynomials. Then the order of the $x_1, \ldots, x_l$ does not matter in the solutions $X_l$; this should be borne in mind by the reader in what follows. Let $s_j(X_l)$ denote the $j$-th elementary symmetric polynomial of $x_1, \ldots, x_l$. The special system of equations of the form

$$s_j(X_l)=\sigma_j \qquad (\sigma_j \in \mathsf{F};\ j \in M) \tag{2}$$

is called *elementary symmetric*, where $M$ may be any subset of $\{1, \ldots, l\}$. If all the $\sigma_j$ are equal to 0, the system (2) is said to be *homogeneous*, while it is called *complete* if $M=\{1, \ldots, l\}$ holds. A complete system has, of course, at most one solution.

Elementary symmetric systems always play a role in connection with a symmetric system (1) as the latter can be written (uniquely) in the form

$$F_i(s_1, \ldots, s_l)=0 \qquad (i=1, \ldots, k), \tag{3}$$

where $s_i$ stands for $s_i(X_l)$, so that the solutions may be obtained in two steps, viz. by first determining the solutions $s_j=\sigma_j$ ($j \in M$) of the "ordinary" system (3) and then those of (2), where $M$ is the set of the suffices of the $s_j$ that actually occur in (3).

But of all this we shall deal here only with the special case

$$s_i(X_l)=0 \qquad (i=1, \ldots, k) \tag{4}$$

for certain $k$, $l$ with $1 \leqq k < l$. We write this homogeneous elementary symmetric system of equations as $S_k(l)$. The case $k=l$ we have disregarded, as $x_1=\cdots=x_l=0$ is then the only solution of (4).

For an arbitrary $X_k \in (\mathsf{F}, \mathsf{F}, \ldots)$ we form the fully reducible polynomial

$$f(x) = \prod_{j=1}^{l} (x - x_j) \quad (\in \underset{1}{\mathsf{F}[x]}) \qquad (f^{\circ} = l). \tag{5}$$

Then (4) and (5) are related with each other by the fact (already considered several times in another form) that $X_l \in (\mathsf{F}, \mathsf{F}, \ldots)$ is a solution of $S_k(l)$ if and only if

$$f^{\downarrow} \geqq k+1. \tag{6}$$

NOTE. Up to now F has been allowed to be any field. If we take an algebraically closed field for F, then (and only then) (2) is always solvable (and in the case $M=\{1, \ldots, l\}$ it has only one solution). In all other cases (2) and $S_k(l)$ lead to interesting but difficult questions. If F is the field of real numbers, it follows from Descartes' rule of signs that even the system consisting of only two "neighbouring" equations

$$s_i(X_l) = 0, \quad s_{i+1}(X_l) = 0 \qquad (1 \leqq i \leqq l-1)$$

has only the trivial solution $x_1 = \cdots = x_l = 0$. In particular, for $i=1$ this follows also from the identity

$$s_1(X_l)^2 - 2s_2(X_l) = x_1^2 + \cdots + x_l^2,$$

but the other values of $i$ also admit a direct proof.

Let us return to the finite field F and consider the case $l=q$. Then we have to deal with the system of equations $S_k(q)$ for a given $k$ such that $1 \leqq k \leqq q-1$. Accordingly, (5) now becomes

$$f(x) = \prod_{j=1}^{q} (x - x_j) \quad (\in \underset{1}{\mathsf{F}[x]}) \qquad (f^{\circ} = q). \tag{7}$$

As $f^{\downarrow} = f^{\circ} - f^{\circ\circ} = q - f^{\circ\circ}$, the following rule arises from the above: $X_q \in (\mathsf{F}, \mathsf{F}, \ldots)$ is a solution of $S_k(q)$ if and only if

$$f^{\circ\circ} \leqq q-1-k. \tag{8}$$

In order to draw further conclusions from this, we define the *multiplicity-number* of an $X_q \in (\mathsf{F}, \mathsf{F}, \ldots)$ as the greatest divisor $P$ of $q$ such that each element of F appears among the terms $x_1, \ldots, x_q$ of the sequence $X_q$ with multiplicity ($\geqq 0$) divisible by $P$. This means in other words, that $X_q$ can be composed of $P$ equal sequences, $P$ being as large as possible.

Let us consider $X_q \in (\mathsf{F}, \mathsf{F}, \ldots)$ with multiplicity-number $P$. We shall disregard the case $P=q$ for the time being. Then, by (7), we have

$$f(x)=g(x)^P \tag{9}$$

with

$$g(x) \underset{1}{\in} \mathsf{F}[x] \setminus \mathsf{F}[x^p], \quad g^{\circ} = \frac{q}{P}, \quad g^{\circ\circ} = \frac{f^{\circ\circ}}{P}. \tag{10}$$

Of course, $g(x)$ is also fully reducible.

The condition (8) is now equivalent to

$$g^{\circ\circ} \leqq \frac{q-1-k}{P}. \tag{11}$$

So, from Theorems 17 and 18 (and the note in brackets after Theorem 18) the following statement is valid: Apart from the cases $P=q$ and $f(x) = x^q - x$, $X_q$ is a solution of the system of equations $S_k(q)$ only if

$$\frac{q-1-P}{P} \geqq \frac{q+P}{P(P+1)},$$

i.e.

$$k \leqq q-1-\frac{q+P}{P+1}\left(= q-2-\frac{q-1}{P+1} < q-2\right). \tag{12}$$

We write (12) in the form

$$P \geqq \frac{k+1}{q-k-2}. \tag{13}$$

We have thus obtained that (the inequality $k < q-2$ and) (13) must necessarily hold for the multiplicity-number $P$ of a solution $X_q$ of $S_k(q)$, except for the case when ($P=q$ or $f(x)=f(x) = x^q - x$, i.e.) $X_q$ consists of nothing but equal or all distinct elements of $\mathsf{F}$.

We formulate the result (in full detail) as follows:

THEOREM 30. *Let $P$ be the multiplicity-number of a solution $x_1, \ldots, x_q$ of the homogeneous elementary symmetric system of equations*

$$s_1(x_1, \ldots, x_q)=0, \ldots, s_k(x_1, \ldots, x_q)=0 \qquad (1 \leqq k \leqq q) \tag{14}$$

*(so $P|q$). With the exception of the (trivial) solutions, where the $x_1, \ldots, x_q$ are equal or are all distinct, $k \leqq q-3$ and*

$$P \geqq \frac{k+1}{q-k-2} \tag{15}$$

*must hold.*

COROLLARY. *If* $Q|q$, $Q>1$ *and*

$$k \geqq \frac{Q}{Q+p}(q-1), \tag{16}$$

*then every solutions of the system* (14) *has a multiplicity-number* $P \geqq Q$ *or it consists of all the elements of* $\mathsf{F}$.

The corollary remains to be proved. If $k \geqq q-2$ the corollary follows from the theorem. So let us assume that $k \leqq q-3$. Then it follows from (16) that

$$\frac{k}{q-k-1} \geqq \frac{Q}{p},$$

and so, by (15), we have $P > \frac{Q}{p}$, i.e. $P \geqq Q$. This proves the corollary.

EXAMPLE. For $Q=q=p$ (16) becomes $k \geqq \frac{p-1}{2}$. Therefore the system of equations

$$s_1(x_1, \dots, x_p)=0, \dots, s_{\frac{p-1}{2}}(x_1, \dots, x_p) = 0 \qquad (x_1, \dots, x_p \in \mathsf{F}_p)$$

has only the $p$ solutions

$$x_1 = \dots = x_p = c \qquad (c = 0, \dots, p-1)$$

(of multiplicity-number $p$) and the solution consisting of $0, \dots, p-1$. (Cf. Rédei [6, 13].)

# SOME UNSOLVED PROBLEMS

The problems to be given here are to some extent bound up with the material treated in this book. They are also partly connected with the Hajós factorizations of finite abelian groups.

*Problem 1.* To find the extensions of the applications treated in the text that can be obtained by fuller use of Theorem 17.

*Problem 2.* To find further theorems for lacunary polynomials over $\mathsf{F}$ admitting applications, if possible.

*Problem 3.* Is the conjecture that the number in Theorem 19 takes for every $q$ all integer values in the given intervals correct?

*Problem 4.* To formulate the analogue of Theorem 22 for $\mathsf{F} \supset \mathsf{F}_p$ (i.e. $n \geqq 2$).

*Problem 5.* To determine all the (normed) schlicht decompositions $G = A \cdot B$ of a finite elementary abelian $p$-group of order $p^3$. Conjecture: One of the factors $A$ and $B$ must lie in a maximal subgroup of $G$. [If this is true, then all the possible decompositions of $G$ can be easily determined (with the aid of Theorem 23).] This problem appears to be very difficult.

*Problem 6.* To add to the number of examples given in § 36.

*Problem 7.* To prove the conjecture that in Theorem 24 the case $O(\mathfrak{W}) = \frac{q+1}{2}$ is impossible (also for $n \geqq 2$). (Cf. Theorem 24.) Is $O(\mathfrak{W})$ capable of assuming all the other values given there?

*Problem 8.* To provide generalizations of Theorem 25.

*Problem 9.* To extend the (number-theoretical) applications, made in § 38, in such a way that the case $n \geqq 2$ ($\mathsf{F} \supset \mathsf{F}_p$) of Theorems 5, 17, 18 is also used.

*Problem 10.* To solve the exceptional cases $q = 3^n, 5^n, 7^n$ ($n \geqq 4$) of Theorem 17. We conjecture that in fact these cases are also not exceptions.

*Problem 11.* To try to deduce directly one of the applications of our lacunarity theorems. (In our opinion there is, however, little hope of success here.)

*Problem 12.* To devise new types of applications of our lacunarity theorems.

*Problem 13.* To prove the covering theorem for finite abelian groups (which comes from Hajós's main theorem) (Rédei [1]) directly. (This would be an entirely new proof of this main theorem.)

*Problem 14.* To devise further applications for our zeta-functions (Rédei [7]), going beyond our inertia theorem. To prove directly its pole property, which is equivalent to Hajós's main theorem (Rédei [8]). (This, too, would be an entirely new proof of this main theorem.)

*Problem 15.* To find for a fixed $\mathsf{F}$ the greatest natural number $k\left(<\frac{q}{2}\right)$ such that of the polynomials of the form

$$f(x) = x^q + \alpha x^{q-1} + \cdots + \alpha_k x^{q-k} + \cdots + \beta_k x^k + \beta_{k-1} x^{k-1} + \cdots + \beta_0 (\in \mathsf{F}[x]).$$

only $x^q - x$ is fully reducible.

*Problem 16.* To develop the foundations of a theory of differential equations in $\mathsf{F}[x]$.

*Problem 17.* (P. Turán.) To determine the fully reducible trinomials (of degree $\geqq 4$) in $\mathsf{F}[x]$.

*Problem 18.* To determine, as a generalization of Problem II, for a $k = 2, 3, \ldots$ the fully reducible $f(x)$ such that

$$f(x) \underset{1}{\in} \mathsf{F}[x] \setminus \mathsf{F}[x^p], \quad f^\circ = kq, \quad f^{\circ\circ} \leqq \frac{kq+1}{2}.$$

It is easy to obtain an analogue of Proposition 9. When $p \nmid k$, $(x^q - x)^k$ is a trivial solution. If $2|k$ there are no further solutions. For $2 \nmid k$ all further solutions must belong to the "marginal case" ($p \neq 2$ and) $f^{\circ\circ} = \frac{kq+1}{2}$. If the latter are written in the form

$$f(x) = (x^q - x)^k + \varkappa g(x) \qquad (\varkappa \in \mathsf{F} \setminus 0,\ g(x) \underset{1}{\in} \mathsf{F}[x]),$$

then the differential equation

$$(g(x)^2)' = (x^q - x)^k + \left(\varkappa + \frac{2k}{\varkappa} x^{k-1}\right) g(x)$$

must be satisfied. (Clearly, after the solutions have been derived, applications can be found.)

*Problem 19.* To devise (number-theoretical) applications for the Minkowski–Hajós "marginal" theorem.

# LITERATURE


[1] L. Carlitz, A note on exponential sums, Acta Sci. Math. (Szeged), 21 (1960), p. 135—143.

[2] G. Hajós, Über einfache und mehrfache Bedeckung des $n$-dimensionalen Raumes mit einem Würfelgitter, Math. Z., 47 (1942), p. 427—467.

[3] H. Hasse, Zahlentheorie. Second enlarged edition. Berlin, 1963.

[4] L. Kalmár, On the fundamental theorem of number theory, Mat. Phys. Lapok, 43 (1936), p. 27—45 (in Hungarian).

[5] L. Rédei, Bemerkung zu meiner Arbeit „Über die Gleichungen dritten und vierten Grades in endlichen Körpern", Acta Sci. Math. (Szeged), 11 (1947), p. 184—190. (In Hungarian with German abstract.)

[6] L. Rédei, Zwei Lückensätze über Polynome in endlichen Primkörpern mit Anwendung auf die endlichen Abelschen Gruppen und die Gauss'schen Summen. Acta Math., 79 (1947), p. 273—290.

[7] L. Rédei, Zetafunktionen in der Algebra, Acta Math. Acad. Sci. Hung. 6 (1955) 5—25.

[8] L. Rédei, Die gruppentheoretischen Zetafunktionen und der Satz von Hajós, Acta Math. Acad. Sci. Hung. 6 (1955), p. 271—279.

[9] L. Rédei, Algebra, Vol, 1, Pergamon Press, Oxford–London, 1967.

[10] L. Rédei—P. Turán, Zur Theorie der algebraischen Gleichungen über endlichen Körpern, Acta Arithmetica 5 (1959), p. 223—225.

[11] L. Rédei, Ein Überdeckungssatz für endliche abelsche Gruppen in Zusammenhang mit dem Hauptsatz von Hajós, Acta Sci. Math. Szeged, 26 (1965), 55—61.

[12] L. Rédei, Die neue Theorie der endlichen abelschen Gruppen und Verallgemeinerung des Hauptsatzes von Hajós, Acta Math. Acad. Sci. Hung., 16 (1965), 329—373.

[13] L. Rédei, Berichtigung zu meiner Arbeit „Die neue Theorie der endlichen abelschen Gruppen und Verallgemeinerung des Hauptsatzes von Hajós", Acta Math. Acad. Sci. Hung., 17 (1966) p. 461.

[14] L. Rédei, Polynome mit eingeengtem Wertevorrat über Körpern, Abhandlungen aus dem Math. Sem. d. Univ. Hamburg, 33 (1969), 29—31.

[15] J. Surányi—P. Turán, Notes on interpolation I. On some interpolatorical properties of the ultraspherical polynomials, Acta Math. Acad. Sci. Hung., 6 (1955), 67—80.

[16] E. Wittman, Einfacher Beweis des Hauptsatzes von Hajós—Rédei für elementare Gruppen von Primzahlquadratordnung, Acta Math. Acad. Sci. Hung., 20 (1969), 227—230.

# SUBJECT INDEX

(Current denominations and those with "local" meaning are not contained in the Subject Index)

# LIST OF THEOREMS, LEMMAS AND PROPOSITIONS